Leading
People-Based Safety™

Enriching Your Cultu

E. Scott Geller, Ph.[

Published by

COASTAL ®

...ustry praise for People-Based Safety™:

One of the many things I have learned over the years as a safety professional is that "safety is a people business." As an attentive learner of Dr. E. Scott Geller for nearly 15 years, I have had the opportunity to practice in the workplace many of the concepts and principles of People-Based Safety.™ People-Based Safety™ allows us to bring the human dynamic into industrial safety. Actively caring has become our value statement at the Idaho National Laboratory. People-Based Safety™ works!

Bowen W. Huntsman,
Advisory Engineer, Voluntary Protection Program, Idaho National Laboratory

As a motivational safety speaker, I have the privilege of speaking for hundreds of clients. I can tell you the ones who use People-Based Safety™ are thrilled with their results, and I can see their performance stands way above the crowd. Their employees show a real caring about the safety of others on and off the job.

John Drebinger, Certified Speaking Professional

The Idaho National Laboratory (INL) has embraced the concepts of People-Based Safety™ as presented by Dr. E. Scott Geller. The concepts and principles of People-Based Safety™ have been an integral component in the laboratory's being recognized as a Department of Energy Voluntary Protection Program (DOE-VPP) STAR site. Our employee observation and feedback process is entitled SOAR (Safety Observations Achieve Results). SOAR is a significant continuous improvement step toward world-leading safety within industry and the DOE Complex. It is advancing us, through a greater focus on workplace observations, into a stronger sense of responsibility toward protecting ourselves and our coworkers. It truly is a "people-based" approach to safety that has resulted in employees actively caring for each other in a positive safety environment.

Ronald K. Farris, Human Performance Specialist, Battelle Energy Alliance

PBS at Bechtel is not only a worker observation/feedback process, it is also evolving into total engagement at all levels of the organization, looking at and focusing on the ES&H behaviors that continue to drive our company to zero incidents, and improving on our world-class ES&H culture.

Bill "Duke" Luksis, Bechtel Environmental, Safety & Health Services

People-Based Safety™ has been a cornerstone of Ensign's journey to zero incidents. Because this is an employee-led process based on valuing and respecting each individual's perspective of the job, employees have steadily and considerably improved in participation and reduced the risks they are willing to take. This has translated into significant reductions in the number and severity of incidents, as well as making Ensign a company that people want to work for. Without SPS and People-Based Safety,™ the journey would have been much more challenging and would have taken ten times longer.

Pete Flatten, HSE Manager, ENSIGN Well Services Inc.

People-Based Safety™ has dramatically improved the safety, attitudes, and well-being of all of our workforce. It has brought crafts, supervisors, union leaders, and managers together in a true interdependent relationship for the safety of all employees. Prior to system-wide implementation in 2000, we had one fatality every 14 months. Since implementation in 2000, we have gone from fourth to first place in Safety of the Class 1 Railroad Mechanical Departments and have not had a single fatality.

Larry Earnshaw, Director, Safety and Environmental, Union Pacific Railroad Company

Feedback received from People-Based Safety™ is the foundation for self-improvement. This dynamic process improves interpersonal interactions. Leading People-Based Safety™ creates more building blocks for culture enrichment. As quoted from Proverbs 29:18, "Where there is no vision, the people perish." The vision for People-Based Safety™ is a community of people actively caring for the well-being and safety of workers, families and our country.

Joanne Dean, Safety Director, Gale Construction Company, a subsidiary of Mack/Cali

Embracing the concept of People-Based Safety™ empowers organizations to create a work environment that is enriched by trust. Employees feel in control of their jobs, and OHS performance and quality and productivity continuously improve.

Susan Kadar, Director, Lloyd-Jones Meakin Group

People-Based Safety™ presents a holistic approach to the well-being of persons at work. Workplace condition, and systems of work, are of significance, but it is the way people work within those parameters that is of profound importance. People-Based Safety™ goes beyond focusing on behavior and explores why people do the things they do.

Martin Ralph, Managing Director, Industrial Foundation for Accident Prevention

We have been involved in the Total Safety Culture process since 1995. Our [work]site has progressed tremendously using the tools and training Scott, along with the SPS group, first brought to us as "behavior-based safety." The driving force, safety accountability starting at the lowest level, is what brought our attention to the process. Then, moving into "Stop passing the buck" helped us buy in to the culture of People-Based Safety.™ Now, after more than a decade of having ownership of the process, we are looking forward to new tools for *Leading People-Based Safety™*.

Gary Pierce, ExxonMobil Chemical process operator

The principles and strategies of People-Based Safety™ help break the cycle of collusion that creates the victim mentality that is represented in the language, "Us vs. Them." It's a step on a journey to a new culture with an open dialogue that's built on trust and mutual respect.

John Toups, President and CEO, J.W. Toups Inc.

Published by
Coastal Training Technologies Corporation
500 Studio Drive, Virginia Beach, Virginia 23452
757-498-9014 • www.coastal.com

ISBN 978-0-9664604-2-1

Library of Congress Control Number
Provided Upon Request

Printed in the United States of America

Dedication

I started my teaching and research career at Virginia Tech in 1969. For 38 years, I have had the pleasure and honor of sharing profound knowledge with, and receiving gratifying feedback from, thousands of students. Many, many of these young men and women have gone on to make a difference in productive careers and in the lives of countless others.

This book on people-based leadership is dedicated to the memory of 32 university leaders who were killed on April 16, 2007, on the campus of Virginia Tech: Ross Abdallah Alameddine, Christopher James Bishop, Brian Roy Bluhm, Ryan Christopher Clark, Austin Michelle Cloyd, Jocelyne Couture-Nowak, Kevin P. Granata, Matthew Gregory Gwaltney, Caitlin Millar Hammaren, Jeremy Michael Herbstritt, Rachael Elizabeth Hill, Emily Jane Hilscher, Jarrett Lee Lane, Matthew Joseph La Porte, Henry J. Lee, Liviu Librescu, G.V. Loganathan, Partahi Mamora Halomoan Lumbantoruan, Lauren Ashley McCain, Daniel Patrick O'Neil, Juan Ramon Ortiz-Ortiz, Minal Hiralal Panchal, Daniel Alejandro Perez-Cueva, Erin Nicole Peterson, Michael Steven Pohle, Jr., Julia Kathleen Pryde, Mary Karen Read, Reema Joseph Samaha, Waleed Mohamed Shaalan, Leslie Geraldine Sherman, Maxine Shelly Turner, and Nicole White.

The leadership loss from this devastating tragedy is immeasurable and will never be recovered. We can only hope current and future leaders will apply the lessons learned to prevent a similar catastrophe.

E. Scott Geller, Ph.D.
Alumni Distinguished Professor
Department of Psychology
College of Science
Virginia Tech
Blacksburg, Virginia

Contents

Illustrations and Tables

Chapter 5

Chapter 6

Foreword

FairPoint has had a pretty good behavior-based safety program; "accidents" and incidents were down considerably, active participation in local safety programs and projects was up, and our safety program overall was beginning to get "comfortable." It had been four years since our last "major accident" but incidents, close calls and minor accidents were rising in frequency — we had all the makings for "failure-by-positive-apathy."

We began introducing People-Based Safety™ in our "high-incident" target region after an orientation and subsequent buy-in by senior management. Though we have just recently finished sending the first groups through training, the employee-students are actively engaged in Acting, Coaching, Thinking and Seeing (the major tenets of the PBS program) with interest and enthusiasm. They have shown a renewed dedication to safety with fresh attitudes. A great percentage of them have openly commented that PBS is the next logical step in a proactive safety program — one that doesn't just list the safety rules, but engages the active behavioral processes and human dynamics in each of us.

Comments received include:

" . . . presented a different way to look at workplace safety as well as a logical way to approach it."

" . . . empowers each person to be responsible for their own safety as well as their coworkers."

" . . . made me realize what I overlook or take for granted every day."

" . . . People-Based Safety™ makes you realize why you want to be safe."

Kevin Keefe
Corporate Safety Manager
FairPoint Communications, Inc.

For generations our company was guided by this philosophy — that our society, culture, and personal experiences directly impact our work environment. We used that consensus to build a solid safety culture.

But times change. Our society has become less homogeneous; our personal experiences vary more dramatically. And our company realized along the way that the ability to create a safe work environment simply by relying on a long-established sense of common ground had us struggling to maintain the status quo with our safety culture and unable to progress further. Looking to find something new that would energize and transform our safety culture, the People-Based Safety™ program quickly rose to the top as our choice for change for several important reasons:

First, it created a common understanding throughout our culture by defining the safety problems all workers, face regardless of specific job responsibilities.

Second, it created a common language, allowing our company to speak to each other about safety clearly and without misunderstanding.

Third, People-Based Safety™ created common behaviors by teaching sound and practical safety techniques that work, regardless of the environment our employees find themselves in day to day.

We are using People-Based Safety™ as a way to empower our associates to take control of their own safety. Using the tools and techniques taught through the program, every individual within our organization now has both the knowledge to identify hazards of the workplace — and the tools and techniques necessary to work safely.

People-Based Safety™ has pointed our company toward a unified safety culture, bound together by common definitions, behaviors, beliefs, tools and techniques, and a common sense of purpose. It is a culture focused less on rules and more on attitudes, where safety is a part of our moment-to-moment thought processes rather than an afterthought. And it is a culture where safety is viewed as integral to the company's long-term health and not as an obstacle to company success.

Ian M. Cox
Engineering & Operations Coordinator
Turkey Hill Dairy

Preface

Leadership books are plentiful — I count 67 in my office alone — and they overlap greatly regarding the qualities and behaviors of effective leaders. Most of these books are read by corporate managers and supervisors to get more, better, deeper, faster, more authentic (you get the idea) effort from their followers.

Leading People-Based Safety™ is for everyone. And it includes principles and procedures you can apply beyond safety — to build trust and caring in your culture, mindfulness and self-direction in your coworkers, and improve the overall quality of conversations, relationships, coaching and risk perceptions in your organization.

Everyone can be a leader, even and perhaps especially when you are working alone. We all encounter opportunities daily — to listen, ask questions, empathize, compliment or correct — to extract the best from ourselves, our coworkers, our families, and our organization's culture.

Managers hold us accountable to complete certain tasks; leaders inspire us to be self-accountable — to do right because we want to, not because we are told to, we are being watched, or because the rules say so.

Leaders help people become self-directed about their safety and many other competent behaviors. At whatever level you are positioned in an organization, you have a leadership role to help those around you perceive personal ownership of goals, new initiatives and assignments. You can contribute to people wanting to improve for the team — their coworkers — and take personal pride in their accomplishments. And you can bring an "actively caring" mindset to your daily conversations and interactions — from solving problems and giving corrective feedback to providing the kind of supportive feedback that feels rewarding and boosts the person states, or states of mind, that enrich a unified, actively-caring culture.

This book should be distributed throughout your entire organization. The many lessons of human dynamics can benefit interpersonal communication, build win/win trust, cultivate synergistic relationships, enrich a work culture, and maximize the safety of a workforce. My hope is readers will grasp the principles of ACTS — Acting, Coaching, Thinking and Seeing — and be inspired to use them at work and at home. Original illustrations by George Wills are interspersed throughout to portray certain lessons (all derived from evidence-based behavioral science) and add some relevant humor.

No cookbook

It is impossible, though, to provide you step-by-step procedures, some sort of cookbook for leadership. You will need to customize these assorted principles from "humanistic behaviorism" for your particular situations —

your unique culture. This enhancement happens when you and other partic-ipants in your workplace learn and believe the principles, discuss possible applications, and then use various aspects of Acting, Coaching, Thinking and Seeing to enrich yourself, your team, and your culture.

Leading People-Based Safety™ extends and refines the content in *People-Based Safety*™: *The Source* (Coastal Training Technologies Corporation, 2005) in order to offer advice for leadership. Here's the bottom line: the principles of PBS cannot make a difference in the work and lives of people, and in orga-nizational cultures, without leadership that activates the acceptance of personal responsibility to customize and implement all that PBS offers.

In addition to the leadership theme, this book is most unlike *People-Based Safety*™: *The Source* in its expanded discussion of personality. Several aspects or dimensions of personality are introduced here, together with tools for esti-mating personal variation along certain characteristics. This enables you to more profoundly appreciate individual differences, and increases your under-standing of personal and interpersonal strengths and weaknesses. The result: improved safety performance, conflict resolution, and relationship-building for culture enrichment.

Indeed, all of this book's content — from becoming a world-class leader to improving communication and understanding personality — will help you evaluate your own personal potential, develop constructive relationships, and cultivate a Total Safety Culture. Your challenge is to study and apply these evidence-based principles of human dynamics. I urge you to think big. Envision a plethora of positive consequences emanating from a culture enriched by the PBS leadership principles explained in this book. Then you will be motivated to take action.

E. Scott Geller
June 2007

Acknowledgments

*L*eading People-Based Safety™: Enriching Your Culture evolved from my most recent contributions to *Industrial Safety and Hygiene News* (ISHN). Since 1990, I have written a monthly article for this publication's column on "The Psychology of Safety." Before submitting each essay to the editor of ISHN, I email the draft to a group of colleagues and solicit their feedback. Over the years, the amount of feedback participation per individual group member has varied considerably.

The select group of individuals who provided consistent and constructive feedback from the scholarship on which this book is founded include Susan Bixler, Tommy Cunningham, Joanne Dean, Leah Farrell, David Harris, Phil Lehman, Anne Lewis, Tim Ludwig, Steve Roberts, Bob Veazie, Doug Wiegand, and Josh Williams. These are leaders who made a special effort to read rough drafts of my monthly essays over the past three years and offered thoughtful commentary — thank you so very much.

After my revised article is submitted to ISHN, readability is greatly improved by the talent and insight of Dave Johnson — the editor of ISHN. Indeed, Dave and I have collaborated on disseminating principles of psychology applied to occupational safety for almost 20 years. He was the editor of my first safety book — *The Psychology of Safety* (Chilton Publishers, 1996) — as well as this text. And we recently co-authored a book that adapts and teaches the psychology of safety for healthcare workers — *People-Based Patient Safety™: Enriching Your Culture to Prevent Medical Error*. Clearly, no one knows my perspective on the psychology of injury prevention, including relevant leadership and culture-change principles, better than Dave Johnson.

Dave Johnson introduces each of the six chapters of this text, as he did for *People-Based Safety™: The Source*. As editor of my scholarship, Dave has helped me immeasurably in making the human dynamics of safety leadership real — clear, coherent, and applicable. I am eternally beholden to Dave Johnson.

Because the first drafts of my scholarship are handwritten, I am always indebted to someone's word-processing skills. For this book, I gratefully acknowledge Christina Goodwin and David Harris — the 2006–2007 coordinators of our Center for Applied Behavior Systems in the Psychology Department at Virginia Tech. In addition, David Harris organized the text and illustrations for the publisher.

Each of my safety-related texts and workbooks benefit from the extraordinary talent of George Wills, the creator of the instructive and entertaining illustrations. This text includes a number of original cartoons by this artist that have not appeared elsewhere. I hope you appreciate George's illustration skills as much as I.

Finally, I thankfully acknowledge the dedication and continual support of Coastal Training Technologies Corporation, particularly the leadership of Nancy Kondas, Marshall McClure, and Terry Wygant. Thank you, Nancy, for envisioning this text and then guiding each step toward publication. Thank you, Marshall, for creating the design of this book, from cover to cover. And, thank you, Terry, for promoting the concept of a People-Based Safety™ leadership book, both before and after the publication of this text.

I have been blessed with an extensive support system both in the academic and consulting worlds — professional colleagues, university students, and consumers of my books and education/training programs. All of you have offered invaluable feedback to help me improve, and you have inspired me to keep researching and developing ways to benefit the human dynamics of injury prevention, behavioral intervention, and culture enrichment.

E. Scott Geller
June 2007

CHAPTER ONE

Keeping People-Based Safety™ Going

Introduction

In the two years since People-Based Safety™ was introduced, perhaps the most common question to come from users of the PBS process is this: "Dr. Geller, now that we're on our way, how do we take our PBS program to the next level?"

This book answers that question. Leadership is required to truly embed the principles of People-Based Safety™ in your culture for the long run. But Dr. Geller is not talking here about traditional top-down leadership, with distant managers decreeing policies, procedures and protocols — anything but.

For one thing, there is no single way to "do" PBS, to take it to the next level. There is no cookbook. This is one distinction between People-Based Safety™ and Behavior-Based Safety, with which many of you are familiar.

Indeed, there is something of a step-by-step methodology to BBS. Dr. Geller captured it years ago in his "DO IT" model. You define your organization's critical safety-related behaviors. Itemize them on a checklist and go out and observe them. Intervene by noting if the observed behavior is safe or at-risk and offer feedback. Finally, you track the overall percentage of observed safe versus at-risk behaviors, and you celebrate — recognize your group's success — as the percentage of at-risk behaviors drops and the percentage of safe behaviors rises.

The "DO IT" process is reviewed here in Chapter One, right at the start in Part 1. It effectively transfers from BBS to become one of the basics of PBS. But as users of PBS have learned, there is much, much more to PBS than the 30-year-old practice of workplace observation and feedback. Five distinctions between PBS and BBS are discussed in Part 2 of this chapter.

The rest of Chapter One presents a primer of sorts on People-Based Safety.™ You can't lead PBS, use it to enrich your culture, and take your safety efforts to the next level if you don't grasp the essential tools or skill sets — PBS's core principles of ACTS: Acting, Coaching, Thinking and Seeing.

Part 3 on *Acting* emphasizes 10 strengths of the behavioral approach. As Dr. Geller says, it all begins with behavior, what can be observed.

Part 4 on *Coaching* makes the point that, as Dr. Geller says, "the human dynamics of safety are more complex than mere checks on a behavioral checklist." Everyone coaches in the PBS process, just as everyone takes the lead in his or her own way.

This sheds light on why there is no single way to implement PBS. Everyone coaches differently. Everyone leads differently. The underlying principles of coaching and leadership are constant, but how you haul them out and hone them to your liking depends on variables such as your work environment, the makeup of your workforce, and those attributes that make your culture unique.

Part 5 on *Thinking* discusses important PBS concepts such as mindfulness, self-direction and self-accountability. Again, the nature of the work accomplished in your organization, the number of "lone workers" you employ, for instance, will determine how you teach and support mindful thinking and self-talk.

Finally, Part 6 on *Seeing* reviews six powerful perceptions that influence safety-related behavior, such as: Familiarity with our daily routine breeds contempt for risks.

Leaders need to see that everyone in their workplace owns widely varying views on safety, and has different personalities that affect their perceptions, actions and beliefs. Optimistic, proactive leaders empathize with this diversity and tap into it to construct unique PBS strategies. They don't try to steamroll and flatten diversity with a one-size-fits-all, command-and-control process. To take safety to the next level and truly enrich your culture, you need to get beyond that narrow-minded thinking and be more creative and resourceful.

The remaining five chapters of *Leading People-Based Safety*™ will activate that creativity and resourcefulness.

Dave Johnson, Editor
Industrial Safety & Hygiene News

1 Seven Basics of People-Based Safety™

Behavior modification. . . safety management. . . attitude adjustment. . . behavior-based safety. . . culture change. . . cognitive alignment. . . person-based safety. . . human engineering. . . social influence. Whew! All these terms address the human dynamics of injury prevention.

And most are insufficient. They are either too narrow and restricting in their focus or scope, or too broad and vague. Some focus solely on behavior change; others try to target vague and unobservable aspects of people, like their inner attitudes and thoughts. Still others have the grandiose notion of directly targeting culture change.

All these approaches are well-intentioned — and none is entirely wrong. Indeed, the human dynamics of an organization include behaviors, attitudes, cognitions, and the context (or culture) in which these aspects of humans occur. But some approaches are too equivocal or ambiguous to be practical, while others may be practical but are not sufficiently comprehensive.

An integrated approach

In the early 1990s, I proposed addressing both behavior-based and person-based factors to improve — and sustain gains in — workplace safety. I called this approach "people-based safety" and proposed substituting empowerment, ownership, and interpersonal trust for more traditional safety jargon like top-down control, compliance, and enforcement. My partners at Safety Performance Solutions began implementing these new people-oriented procedures in 1995 under the popular label at the time: "behavior-based safety."

Today we call this approach or process People-Based Safety™ (PBS). It strategically integrates the best of behavior-based and person-based safety to enrich workplace cultures — improving job satisfaction, quality and production, interpersonal relationships, and, of course, occupational safety and health.

To begin this book on *Leading People-Based Safety™: Enriching Your Culture*, I want to give you a primer on the basics of PBS. Before you can lead with PBS, you need to grasp the essential principles and procedures. Here I introduce seven underlying principles of PBS.

Principle 1: Start with observable behavior

Like behavior-based safety, PBS focuses on what people do, analyzes why, and then applies research-supported interventions. Safety improvements and other positive outcomes result from *acting people into thinking differently* — rather than targeting internal awareness or attitudes in order to *think people into acting differently*.

But here is a significant distinction: unlike behavior-based safety, PBS considers that people can observe their own thoughts and attitudes. People can think themselves into safer actions. This self-management requires self-dialogue or thinking as well as self-directed behavior.

Principle 2: Look for external and internal factors to improve behavior

A behavior analysis of work practices will pinpoint many external factors that encourage at-risk behavior and hinder safe behavior. But here's another distinction: it's also possible for individuals to self-evaluate their own self-talk and selective perceptions regarding safety-related behavior, and choose to make appropriate adjustments. PBS teaches people how to address their internal thoughts, perceptions, and attitudes.

Principle 3: Direct with activators and motivate with consequences

Activators (signals or events preceding behavior) are only as powerful as the consequences supporting the behavior.[1] Activators tell us what to do in order to receive a pleasant consequence or avoid an unpleasant consequence. This reflects the ABC model, with "A" for activator, "B" for behavior, and "C" for consequence. This principle is used to design interventions for improving behavior at individual, group, and organizational levels.

Principle 4: Focus on positive consequences to motivate behavior

Perceptions of personal freedom and responsibility for safety are reduced by controlling negative consequences[2]. Sadly, the common metric used to rank companies on their safety performance is "total recordable injury rate" (or some such count of losses). This puts people in a reactive mindset of "avoiding failure" rather than "achieving success." PBS provides proactive measures employees can achieve.

How do we increase people's perceptions that they are working to achieve success rather than working to avoid failure? Well, even our verbal behavior directed toward another person — genuine approval or appreciation for a task well done — can influence motivation to increase perceptions of personal freedom and empowerment. Of course, we can't be sure our intervention will have the effect we intended unless we measure the impact.

Principle 5: Apply the scientific method to improve intervention

People's actions can be objectively observed and measured before and after an intervention process is implemented. This application of the scientific method provides critical feedback to build improvement.

The acronym "DO IT" says it all:

D = Define the target action to increase or decrease;

O = Observe the target action during a pre-intervention baseline period to identify natural environmental and interpersonal factors influencing it (see Principle 1) and to set improvement goals;

I = Intervene to change the target action in desired directions; and

T = Test the impact of the intervention procedure by continuing to observe and record the target action during and after the intervention program.

Systematic evaluation of a number of DO IT processes can lead to a body of knowledge worthy of integration into a theory.

Principle 6: Use theory to integrate information

After applying the DO IT process a number of times, you will see distinct consistencies:

- Certain intervention techniques will work better in some situations than others;
- Some individuals will be more successful than others using certain interventions;
- Certain interventions will succeed with some work practices better than others.

Summarize relationships between intervention impact and specific interpersonal or contextual characteristics. This will give you a research-based theory of what is most cost-effective under given circumstances.

DEFINE
behavior(s) to target

OBSERVE
to collect baseline data

INTERVENE
to influence target behavior(s)

TEST
to measure impact of intervention

The DO IT process reflects the evidence-based approach to BBS.

SO, WHAT DID YOUR SUPERVISOR SAY?

Negative consequences lead to negative attitudes.

Principle 7: Consider the internal feelings and attitudes of others

The feelings and attitudes of people are influenced by the type of intervention procedure implemented. The effect of interventions requires careful consideration by those who develop and deliver the intervention. This is the essence of empathic leadership taught by PBS.

The rationale for using more positive than negative consequences to motivate behavior (Principle 4) is based on the different feeling states resulting from using positive versus negative consequences to motivate behavior. Importantly, the way your intervention process is introduced and delivered can increase or decrease perceptions of empowerment, build or destroy interpersonal trust, and facilitate or inhibit interdependent teamwork.

The Seven Essentials of People-Based Safety™

Before you read further, study these fundamentals of PBS:

- Start with observable behavior.
- Look for external and internal factors to improve behavior.
- Direct with activators and motivate with consequences.
- Focus on positive consequences to motivate behavior.
- Apply the scientific method to improve intervention.
- Use theory to integrate information.
- Consider the internal feelings and attitudes of others.

2 Distinguishing People-Based Safety™ from Behavior-Based Safety

Before we go further, it's critical to understand the differences between behavior-based safety, which has been used by safety and health professionals since the 1970s, and People-Based Safety™, which has been widely promoted only since 2005 when my book, *People-Based Safety™: The Source*, was published.[3]

The essentials of People-Based Safety™ are integrated into the model, anacronym, that I call "ACTS." In a Total Safety Culture, people Act to prevent injuries, Coach one another to identify barriers to safe acts and provide constructive behavior-based feedback, Think in ways that activate and support safe behavior, and focus and scan to See hazards.

It's fitting the four fundamentals of People-Based Safety™ (PBS) spell "ACTS" because safety depends upon the energy and actions of people. Also, consider that you lead by going into action. Your actions, the examples you set, your modeling, speak louder than words. You don't have to be an eloquent public speaker to be an everyday leader.

PBS targets attitudes, perceptions and thoughts to improve these "person states," leading to changes in critical behaviors. If behavior or actions don't improve, there is no bottom-line benefit to safety.

Let me be clear: PBS is no substitute for behavior-based safety (BBS). Rather, it extends BBS for greater impact. PBS teaches ways to self-coach and increase self-accountability for safety.

Let's look at five distinctions between PBS and BBS.

1) Self-directed behavior

A BBS observation-and-feedback process initiates and sustains *other-directed behavior*. Workers increase safe behavior and decrease at-risk behavior because others — their peers — hold them accountable.

But people often work alone, so they need to coach themselves. This requires self-accountability and *self-directed* behavior. People need to believe in and own the safe way of doing things.

Self-direction requires an internal justification for the right behavior. This happens when external consequences supporting an action are not sufficient to totally justify the behavior. Too often, people choose safe over at-risk acts only because they want to obtain a reward or avoid a penalty. These programs often get the desired behavior — as long as this accountability system is in place. But what happens when the external rewards or penalties are unavailable?

Don't over-justify safe behavior with large incentives and severe threats. Instead, provide education, training, and experience to help people develop a sense of personal control over preventing injuries.[4]

Lone workers need self-direction for safety.

2) Mental awareness

Developing safe habits is a key objective of BBS. Daily repetition of an observation-and-feedback process builds "habit strength," eventually resulting in the development of safe habits. This is good, but not ideal. Habits occur without mental awareness or thoughts, as when you buckle your safety belt without thinking about it before you drive off.

But what if your buckle-up behavior is so automatic you don't notice a passenger in your car is not buckled up? You could miss an opportunity to actively care for the safety of others. And you miss an opportunity to develop self-talk or thinking that supports self-direction and self-accountability.

I hope you agree self-directed and mindful behavior is more desirable than mindless, habitual behavior.

3) **Personal choice**

I've heard many BBS trainers, consultants and students claim that certain environmental cues "trigger" or control safe behavior. This implies that some sort of stimuli cause safety-related behavior to occur. Not true.

Some "triggers" cause involuntary behavior. The flashing blue lights of a state trooper elicit certain emotional reactions. But drivers choose to slow down and pull over. Obviously, traffic lights do not always trigger or control safe intersection behavior. They may cause an emotional rush following a driver's decision to speed through an intersection as the light changes from yellow to red.

Bottom line: There is a space between the stimulus (or activator) and

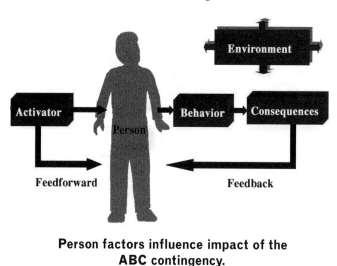

Person factors influence impact of the ABC contingency.

voluntary behavior. Activators provide direction, but it's up to you whether to follow the direction. Your choice is largely determined by how you perceive consequences with regard to their importance to you. What positive consequence do you expect to gain and/or what negative consequence do you expect to avoid? You might choose to run a yellow traffic light because of the beneficial consequences of getting your child to school on time.

Yes, this is the standard ABC (Activator – Behavior – Consequence) Principle of BBS, but the PBS view of it takes into account one's beliefs, perceptions, and attitude.

4) **Attitudes and perceptions**

"Positive reinforcement" is overused and abused by trainers and students of BBS. A consequence is a reinforcer (positive or negative) *only* if it increases the behavior it *follows*.

Attitudes and perceptions determine the motivating potential of a reward or penalty. Trainers and students of PBS realize the reinforcing power of a consequence is in the eye of the beholder. The meaning of a "safety trinket" to an individual determines whether such a consequence is viewed as positive, negative, or neutral, and could motivate behavior.[5]

It is usually impossible to determine whether delivering a consequence actually influences the behavior it follows. Thus, the loose use of "positive reinforcement" among BBS consultants and students is risky and often inappropriate. PBS does not make this mistake. "Positive reinforcement" is not used in PBS, and the impact of positive consequences on feelings or person states is entertained and appreciated.

5) Feeling states

In PBS, positive consequences are considered "rewards," and negative consequences are "penalties." If these consequences don't impact overt behavior, they will at least influence feeling states. Feelings, attitudes, and emotions are important considerations in PBS.

With PBS, rewards increase self-esteem and perceptions of personal competence and control, as well as improve behavior. Research shows these feelings increase people's willingness to actively care for the safety and health of others. Thus, PBS applications of the ABC Principle are directed to both external behaviors and internal person states.

③ People-Based Safety™ — Acting

The remainder of this chapter briefly summarizes the individual components of ACTS — Acting, Coaching, Thinking and Seeing — as part of our primer on PBS. Here I want to review the strengths of a behavioral approach to industrial safety — the Acting component of the ACTS model.

1) Appreciate the principles

Often the only criticism I receive after giving a workshop or audio-conference on the psychology of safety is the lack of specific procedures someone can use to reap the benefits of my teaching. Although I give real-world examples to illustrate behavioral principles, I do not provide specific details for how to use a tool or method in one's work culture.

My vision is that participants will customize principle-related procedures for their workplace. A consultant can be useful during this knowledge transfer and adaptation process and throughout subsequent training of specific procedural steps. This reflects the distinction between education and training.[6]

Even now, in this list of ten keys to succeeding with the behavioral approach, I am not giving you detailed strategies, but rather the basic principles of behavioral safety. My hope is you will incorporate them into numerous aspects of both your work and home life.

2) Use behavioral language

This principle is fundamental, yet seriously overlooked in so many organizations. Indeed, ambiguous non-behavioral language is used *ad nauseam* in corporate vision and mission statements, management expectations, performance appraisals, interpersonal conversations, and on safety signs displayed throughout a workplace.

Exactly what is "world-class safety"?

How does an organization become "leading-edge" in safety?

How does one "think safety" and bring a "safety attitude" to work?

What is "self-initiative" and "self-accountability" when it comes to safety or any other work challenge?

Let's be realistic: How can you fairly evaluate another person on the common performance-appraisal dimensions of "self-motivation," "enthusiasm," "character," "integrity," "creativity," and "emotional intelligence"?

If you want communication to affect what people do, you need to use behavioral language. Whether setting expectations, recognizing accomplishment, solving interpersonal conflict, or delivering corrective feedback, *specify* what behaviors are implicated.

Don't presume people understand your definition of such emotion-laden labels as "risk taker," "underachiever," "overachiever," "team player," and "safety leader." (I discuss my definition of a "safety leader" in Chapter Two of this book.)

When you provide a list of specific, definable desired and undesired behaviors that reflect your viewpoint, you put others on the same page and offer a prescription for performance improvement.

Proper assessment defines desirable and undesirable behaviors.

3) Connect results with behavior

Most of you have heard the warning, "Keep on doing what you're doing and you'll keep on getting what you're getting." Yet managers and supervisors often discuss performance results without reference to behaviors. As a result, they miss an opportunity to recognize, correct, or direct specific action.

How often have you attended a safety celebration at which an exemplary system outcome is acknowledged — lower lost-time injuries or workers' compensation costs — without any mention of the variety of behaviors contributing to the recognized results?

Bottom line: By linking process behaviors to the observed results of a performance system, you not only clarify your perspective, you offer directives people can follow to meet your expectations.

4) *Appreciate the impact of one behavior*

Jerry Pounds, in his 2005 book, *Praise for Profit*[7], offers real-world examples from his consulting experience to show how changing the quantity or quality of one behavior can make a dramatic difference in system performance. He illustrates the value of asking employees to define one thing they could do differently that would improve their performance output.

Behavioral checklists are so common in applications of BBS, some believe BBS is nothing more than observation and feedback with a behavioral checklist. The checklists can be overly complex, overly simple, or over-used. The BBS process can become a cumbersome and meaningless routine of checking "safe" vs. "at-risk" boxes on a laundry list of behaviors. This is better than nothing because it defines the behavioral components of a work assignment and holds people accountable for selecting safe alternatives.

But how often are relevant workers engaged in redefining behaviors and identifying new behaviors to target? When "percent-safe scores" reach 80 percent or when the process becomes a mostly mindless numbers game, it's time to regroup and consider the principles behind this tool. Ask workers regularly to revisit their behavioral checklists and add or subtract critical behaviors. One behavior change can make a difference.

5) *Apply behavioral accountability*

Using behavioral language is the first step in developing an accountability system for performance improvement. And a behavioral checklist is a tool that enables peer-to-peer accountability. Likewise, accountability is possible whenever coaching, corrective feedback, performance appraisals, or incentive/reward programs are behavior-based. Each behavior-based process can give behavioral directives, measure behavioral results, and give behavioral feedback.

The measurement and feedback tools of BBS are invaluable and should not be taken lightly. They are essential for building skills and enhancing individual

"JOE GETS THE SAFETY PRIZE AGAIN. HE WENT ANOTHER 30 DAYS WITHOUT AN ERROR."

Behavioral accountability catches slackers.

and group competence. Even when we are self-directed, we need to measure our relevant behaviors and benchmark with other people's behaviors. This enables us to assess our progress at a particular endeavor and set the kinds of specific goals that can activate improvement and provide accountability.

6) Make feedback strictly behavioral

It's easier said than done, but it is essential to separate behavior from person factors when giving and receiving feedback.

Corrective feedback is not an indictment of one's personality or an indicator of a character flaw. Feedback must not be related to an individual's attitude, motivation, professional competence, or family history.

Feedback is only about behavior. Yes, responding well to supportive or corrective feedback can lead to an improved attitude, motivation, competence, and even a personality state. But the purpose of feedback is only to pinpoint desirable and/or undesirable behavior. The beneficial outcome of behavioral coaching is maximized when this is realized by those who give and receive feedback.

Incidentally, the common term "constructive criticism" is an oxymoron. How can you criticize and be constructive at the same time? For most people, criticism reflects something negative about a person's attitude, character, or personality — characteristics presumed to influence behavior. But "constructive" implies positive change following the "criticism," and sometimes a change in person factors beyond behavior is requested and expected.

When the focus is on behavior, without any implications of person factors or character flaws, feedback that gives specific direction can be constructive.

7) Learn by observing behaviors

In his classic 1996 text, *Human Competence: Engineering Worthy Performance*,[8] Tom Gilbert emphasized that behavioral observation is key to improving personal competence. Effective people do things differently. Effective managers act differently than ineffective managers. The best teachers demonstrate certain behaviors that average teachers do not.

I've heard several safety pros claim 10–20 percent of their workers contribute to 80–90 percent of their OSHA recordables and lost-time injuries. And safety pros report many employees never get hurt. Behavior makes the difference. Productive workers who never get hurt nor put others at risk emit certain behavioral patterns or best practices.

Individuals can learn how to improve their own behavior by observing others, especially when they use a checklist that specifically defines standards of desirable performance. I'm convinced the remarkable success of behavior-based safety is due more to the learning and accountability aspects of interpersonal observation and feedback than by the "percent-safe" scores and graphs derived from compiling the checks on daily behavioral observation cards.

8) *Examine consequences to understand and change behavior*

The topic of "motivation" can be very complex, involving a variety of person-based unobservables like "personal drive," "intention," "self-esteem," "self-affirmation," "optimism," "need to achieve," "need for person control," and even "free will."

Research supports the validity of these motivational concepts, but their usefulness is limited, especially in work settings. The complexity of motivation is simplified and made practical with a behavioral approach. As defined by this principle, behavior is motivated by consequences. People act to gain pleasant consequences or avoid unpleasant consequences.

Inconvenience and discomfort are negative consequences that inhibit safe behavior, and convenience and efficiency are positive consequences that motivate at-risk behavior. These consequences are natural or intrinsic, in contrast to incentive/reward programs that attempt to motivate various inconvenient or uncomfortable behaviors with extra extrinsic consequences like a financial bonus.

Adding positive consequences to situations in order to motivate certain behavior can be costly. Plus, it can be counter-motivational to discontinue a financial bonus or incentive/reward program.

9) *Promote intrinsic reinforcement*

Many recognition programs include extrinsic rewards — individuals may be selected for special acknowledgment and given praise and a material reward for their noteworthy performance. But this extrinsic approach to recognition is not optimal.

Instead, show interest in what people are doing, and you help them appreciate the intrinsic consequences of this ongoing work behavior. All work produces results, but sometimes the output is not obvious or is taken for granted. When others point out the fruits of our labor, the labor can feel more worthwhile, meaning intrinsic consequences are noted and appreciated.

What are the intrinsic reinforcers of taking extra precautions to do a job safely?

Because injuries are generally rare, workers do not naturally connect injury avoidance to their safe behaviors. Through appropriate interpersonal recognition, the value of safety-related behavior can be realized, including the long-term natural consequences of setting the safe example for others to follow.

Do your employees continually offer practical suggestions for making their workplace safer, either by modifying equipment or environmental conditions or by making safe behavior more convenient? Such safety advice increases in quantity and quality when it affects intrinsic consequences. In other words, when employees see adjustments related to their safety suggestion, they are intrinsically reinforced and more likely to offer more suggestions.

With intrinsic reinforcement, people enjoy their work.

10) *Realize the disadvantages of punishment*

I've decried the use of punishment numerous times. Punishment does work when the undesirable behavior is followed by a soon, certain, and sizable punitive consequence. But it's usually impossible to administer this kind of contingency, especially in a work setting. And even when punishment can be implemented appropriately, it can do more harm than good, as Drs. Jon Bailey and Mary Burch emphasize in their 2006 book, *Thinking Like a Behavior Analyst.*[9]

Punishment promotes aggression, hangs a negative image on the punisher, and can disengage the punished person from an entire work process. Plus, the negative emotions promoted by punishment can spread to other workers and work settings.

In *Praise for Profit*, Jerry Pounds starts his critique of punishment with this memorable quote from ex-GE CEO Jack Welch: "When people make mistakes, the last thing they need is discipline. It's time for encouragement and confidence-building."[10]

When it comes to safety, it's critical to learn from near hits and injuries, whether they result from mistakes or calculated risks. Punishment, incorrectly referred to as "discipline" in industry, severely stifles the type of open and frank conversations needed for this kind of learning.

But when conversation following an undesirable incident is viewed as corrective feedback and leads to observable change in process behaviors and/or environmental conditions, beneficial learning and intrinsic reinforcement are the rewards — increasing the quantity and quality of future safety-improving conversation.

Practice, practice, practice

Most safety leaders have heard these 10 principles before and understand the value of each toward improving the safety performance of any organization. People-based leaders appreciate the relevance of behavioral language, accountability and feedback to situations beyond the workplace.

Still, these principles are not practiced enough in daily work or home life. They are simple enough, certainly intuitive, and supported with plenty of research. Sure, it's a challenge to overcome traditions of non-behavioral language in performance appraisals and other accountability systems, but the outcome is well worth the effort — increasing the competence of individuals and groups in any setting and improving the output of any system.

Principles of the Behavioral Approach (Acting in ACTS)

- Appreciate the principles.
- Use behavioral language.
- Connect results with behavior.
- Appreciate the impact of one behavior.
- Apply behavioral accountability.
- Make feedback strictly behavioral.
- Learn by observing behaviors.
- Examine consequences to understand and change behavior.
- Promote intrinsic reinforcement.
- Realize the disadvantages of punishment.

4 People-Based Safety™ – Coaching

As I mentioned previously, the human dynamics of safety are more complex than mere checks on a behavioral checklist. For years managers have used behavioral data to hold people accountable. People-Based Safety™ (PBS) extends the standard BBS process to inspire people to hold themselves accountable.

How?

PBS puts special focus on the actively-caring process of interpersonal safety coaching.

Everyone coaches

You can obtain sufficient upstream behavioral numbers by assigning coaching duties to a select sample of a workforce. Some BBS consultants advocate training a small percentage (perhaps 10 percent) of line employees to be safety coaches. This can save both time and money and is "sold" on the appearance of efficiency.

It's also easier and more efficient to exclude management from the coaching process. A number of BBS consultants train only the hourly workforce to conduct behavioral observation and feedback sessions.

But these kinds of shortcuts limit development of self-accountability and can have only short-term benefits — reflected in the common "pencil-whipping" label given to these types of BBS programs.

People-Based Safety™ presents a broader vision; everyone — salary and hourly workers alike — coaches for safety. Everyone learns the principles and procedures of behavior-based observation and feedback.

Why?

Because coaching develops the self-directed accountability needed for long-term impact. Coaches feel obligated to adopt the principles and procedures they teach and advocate. To be sure, it might be necessary to start a PBS coaching process with a select number of workers. It depends on your work culture, especially the level of interpersonal trust that exists. But it's expected that eventually all employees will coach each other for injury prevention.

Coaching can be informal

PBS teaches and advocates both "formal" and "informal" coaching. Formal coaching parallels the standard BBS application of a critical behavior checklist. Informal coaching involves brief personal conversations to maintain daily attention to safe and risky behaviors and conditions throughout a workplace.

Perception can vary dramatically between individuals.

By focusing more on the process than on checklists and numbers, PBS increases the quantity and quality of informal coaching. This leads to self-coaching, an essential process for the safety of lone workers.

Empathy is essential

In addition to informal conversations, empathy plays a critical role in PBS — from empathic listening to empathic leadership. The Platinum Rule[11] — "Treat others as *they* want to be treated"— is especially pertinent for PBS coaching.

As with client-centered or humanistic therapy,[12] the focus is on the perceptions and feelings of the individual being coached. Behavior and environmental conditions are observed from this person's perspective, and feedback communication is supportive and nondirective. By this I mean feedback is not delivered to direct specific behavior change (as in BBS), but to empower personal responsibility and self-accountability to improve.

It might be efficient to assume people want the same things you want. But it's more effective to discover other people's needs and perceptions before choosing a treatment or intervention approach. Even when the eventual tactic is the same as you would have selected, because you asked first, you can expect greater acceptance, appreciation, and ownership.

This is a brief overview of the coaching component of PBS. I've given particular emphasis to certain qualities that distinguish it from standard BBS coaching. Remember, as I remarked earlier, PBS principles and procedures evolved from BBS, and should be viewed as an extension of BBS, rather than a substitution.

Beyond Observation and Feedback

Unique features of PBS coaching can make any observation and
feedback process more effective. They include:
- Focus on process over numbers;
- Aim to get everyone coaching, including supervisors;
- Promote both formal and informal coaching;
- Develop self-coaching through self-talk and self-accountability;
- Respect the underlying philosophy of empathy.

5 People-Based Safety™ – Thinking

At workshops and keynote addresses on People-Based Safety™ (PBS), I often ask the audience whether they buckle their safety belts automatically, without thinking. Most raise their hands to affirm their buckle-up habit for safety. My reaction: That's good, but not great. It would be better to *think about* what you're doing while fastening your safety belt.

Conscious competence is usually better than unconscious competence, especially when the behavior is safety-related. I'd like to convince you of the validity of this perspective, which deviates markedly from the philosophy of behavior-based safety (BBS). Specifically, BBS promotes development of safe habits as a primary objective of applying BBS tools.

Thinking safe behavior

Thinking is self-talk or internal verbal behavior. I advise my audiences to tell themselves what they are doing when they perform a safety-related behavior. For the safety-belt example, I recommend self-talk that acknowledges the behavior — "I'm buckling up for safety."

"Sir, please re-enter the gate. You're not thinking about workplace safety."

Thinking affects behavior and vice versa.

When safe behavior is accomplished for positive consequences, it is beneficial to also verbalize the rationale for the behavior. What are your personal reasons for choosing safe behavior? For safety-belt use, you might say to yourself, "I'm buckling up to do the right thing for safety — to be a competent driver" or "I'm buckling up to set the safe example for other passengers in my vehicle, and for anyone else who might see me driving."

It's possible, however, your safe behavior is not self-directed, but other-directed. In other words, you might be working safely because someone other than yourself is holding you accountable. For example, some might buckle up to avoid a fine, as suggested by the popular U.S. slogan: "Click it or ticket."

If your safe behavior is other-directed, your self-talk should not include those external controls influencing your behavior. Until you can give a self-directed rationale, you should only tell yourself you are performing the behavior. Forget the external, other-directed reasons for your safe behavior. Let me explain why.

Self-direction and self-accountability

As I reviewed in Chapter Four in *People-Based Safety™: The Source*[13], when people are mindful about their behavior, they are more likely to avoid human error. Self-talk enables the adjustment of behavior per situational factors. It could call your attention to other people not following your safe example, such as a passenger in your vehicle who is not buckled up.

This behavior-based self-talk increases your awareness of the best way to perform under certain circumstances. But there is a more profound reason for thinking about your safe behavior. Your self-talk influences self-persuasion, which in turn enhances self-accountability for safety. Indeed, we hold ourselves accountable by talking to ourselves. What kind of safety self-talk builds our self-accountability or responsibility for safety?

Outside vs. inside control

The reasons we give for our behavior determine the degree to which our behavior is other-directed or self-directed — whether we are accountable to others or accountable to ourselves. This is not an all-or-none state. We can be motivated by both outside and inside controls. But the more our behavior is directed and motivated from within ourselves, the more apt we are to perform the behavior when alone and only accountable to ourselves.

This, clearly, is the ideal safety state for the lone worker.

How do we put ourselves in this state? You know the answer — self-talk. We talk ourselves into being self-accountable. Some situations facilitate this thinking; some do not.[14] In general, self-accountability thinking decreases as the degree of external negative control increases (as in severe threats and strong enforcement), and people's perception of personal choice decreases. Also, the more a behavior aligns with our perception of who we are — our core values — the greater the self-accountability for that behavior.

Self-perception, personal values and self-accountability

Who do you think you are? In other words, what kind of person are you? Do you hold safety as a core value? How do you know?

Our behavior defines us. Each of us is the kind of person who does the things we do. But there are exceptions. When we feel our behavior is controlled entirely by external factors, we do not view that behavior as a reflection of who we are.

When we perceive our behavior as self-directed, we use that behavior to define our attitudes and values.[15] In other words, the behaviors we choose to perform provide information for our self-perception. These behaviors are certainly motivated by expected consequences, both intrinsic and extrinsic. The key is to perceive some degree of choice, and perception of choice is stifled by enforcement or negative reinforcement contingencies (as when we act in a certain way to avoid a negative consequence).[16]

Thus, our self-directed behavior informs our self-perception and our core values. And our self-perception and personal values influence our behavior. We strive for our behavior to be consistent with our values, and vice versa. When we perceive an inconsistency between behavior and the values that define us, we experience tension or "cognitive dissonance" (the academic label used by the many social psychologists who researched this phenomenon). We direct our self-talk to reduce this negative state.[17]

Bottom line: The rationale we provide ourselves for performing safe behavior determines whether we feel self-accountable and will continue to perform that behavior in the absence of an external accountability system. And, of course, the rationale for our behavior is determined by our thinking or self-talk. PBS teaches the kind of thinking needed to develop self-accountability, as well as the kinds of environmental/management conditions/systems needed to promote and support self-accountability thinking.

How Minds Turn On to Safety

- Self-talk, personal reminders, mental notes enable us to monitor and direct our behavior.
- Thinking about our safe behavior helps persuade us that we're doing the right thing.
- Being personally convinced that acting safely is in line with our values and reflects the kind of person we want to be, leads us to hold ourselves accountable for our own safety.
- But this kind of personally motivated thinking fades in the face of severe threats and strong enforcement. Our perception of personal choice decreases. We feel controlled, manipulated.
- The key is to perceive some degree of *choice* in how we act.
- And this perception needs to be supported by personal reasons for acting safe.
- Positive personal perceptions and positive beliefs form a powerful kind of thinking that benefits safety.

6 People-Based Safety™ – Seeing

Do you follow the Golden Rule? Do you "treat others as you want to be treated"?

If so, that's good but not great.

People-Based Safety™ (PBS) teaches the Platinum Rule[18] — "Treat others as *they* want to be treated." We need to understand the perceptions of others before making interventions that impact their lives.

To illustrate this principle, I tell audiences of a memorable experience I had in fifth grade. My teacher called me to the front of the class to be recognized for the superb job I did on my homework assignment. Later, several classmates beat me up on the playground. I didn't want public recognition in the classroom. But my teacher didn't *see* the classroom situation as I did.

Understanding workers' perceptions is a critical challenge of PBS, both when developing and delivering a process to support safe behavior and/or to correct at-risk behavior.

Surveying perceptions

Perception surveys are a useful way to do this. Pre-intervention surveys inform the design of intervention strategies, and comparisons of pre- and post-intervention surveys estimate the diverse impact of an intervention on people's perceptions, attitudes, and values.

My partners at Safety Performance Solutions have been applying the same comprehensive perception survey for more than a decade, and thus have a database of more than 8.5 million safety-related perceptions across a broad range of industries worldwide. These culture surveys are invaluable for benchmarking and for customizing intervention strategies for various types of operations.

Public recognition is not necessarily rewarding.

Bottom line: People's views of safety-related issues vary widely and should be considered when you design and evaluate procedures for improving safety performance. Here are six powerful perceptions that influence safety-related behavior.

1) Familiarity breeds contempt (for risks)

High on the list of psychological factors influencing an individual's perception of risk and safety-related behavior is the role of familiarity. The more experience we have regarding a potential risk, the less risk we perceive.

You can appreciate this principle by recalling your safety-related behaviors when you first started to drive and comparing them with your current driving. As experience bolsters our perception of control, it also increases the possibility of risk-taking.

2) Choice matters

In *People-Based Safety™: The Source*, I discuss the importance of perceived choice when transitioning from other-directed accountability to self-directed responsibility.[19] Consider how less risky those hazards we choose to experience seem (on the road, in the workplace, and during recreation) compared with those hazards we feel *compelled* to endure (like food preservatives, environmental pollution, and earthquakes).

Assessment Tools

We cannot assume other people see what we see and interpret
risks and risk taking as we do. PBS teaches empathy — the need
to assess the perceptions of a work group or culture targeted for
an injury-prevention intervention. These perceptions are carefully
considered when customizing an intervention process.

PBS also teaches techniques for hazard recognition and risk assessment. These include:

- Appropriate alternating between focusing and scanning;
- The Exposure-Severity-Probability (ESP) approach to risk
  assessment;
- The development and application of a Critical Behavior
  Checklist.

These tools are explained in *People-Based Safety™: The Source.*

3) "Safety" in numbers

Perceptions of risk are boosted more readily through the vivid use of individual case examples rather than through use of group statistics. Safety
meetings and interventions should focus on individual experiences rather than

**Personal protective equipment
can influence risky behavior.**

cold, hard numbers. After all, we
can "hide" behind numbers. But
when people hear personal
perceptions and the emotional
regrets of coworkers injured on
the job, they imagine themselves
in a similar unfortunate circumstance. Their perception of risk is
enhanced, and safe behavior
increases.

4) Just desserts

The perception that "people
get what they deserve" has
intriguing implications for workplace safety. I believe this view
contributes to the common
perspective, "It won't happen to
me." Since most believe they are
essentially good and therefore undeserving of a bad-luck injury, they expect

the "other guy" to get hurt on the job — not them. Everyday experiences usually support this perception. Injuries do happen, but not to most individuals, even when they take risks.

5) Superman complex

When you perceive you are well-protected, do you take more risks? Many people do. The implication of this phenomenon is that making a job safer with machine guards or personal protective equipment lowers people's risk perception[20] — and increases at-risk behavior. When safety guards or PPE are added to a work task, behavioral observers should be alert to the possibility of extra risk-taking related to the behaviors protected by the new safety equipment.

But the increase in risk-taking and injuries does not negate the benefits of the protection. Although football players increase at-risk behaviors when suited up with helmets and pads, they sustain far fewer injuries than they would without the PPE.

6) Driven to distraction

I'd like to review one more safety-related perception, one that contributes to many injuries. Specifically, my Type-A personality and need-to-achieve attitude facilitate a future-oriented mindset that gives too much attention to the future and too little to the present. I can still hear my mother admonishing me to "stop and smell the roses."

Perceiving and seizing the moment means being mindful and attentive to our ongoing behavior. We use all relevant senses to recognize what we are doing and where we are doing it. Our antennae are fully extended, enabling us to fully encounter the present. Procedures and tools of PBS help to initiate and support present-focused perceptions and mindfulness.

Essentials of People-Based Safety™

Act to prevent injuries.

Coach one another to identify barriers to safe acts and provide constructive behavior-based feedback.

Think in ways that activate and support safe behavior.

Focus and scan to *See* hazards.

These four essentials of People-Based Safety™ — called "ACTS" — provide knowledge, skills and tools to fully address the human dynamics of workplace safety.

Becoming a World-Class Leader

Introduction

Dr. Geller relates the story at the start of Chapter Two of the frustrated CEO who dislikes the ambiguity of the term "world-class safety." "Everyone talks about it, but no one defines it," he says.

Many safety and health pros share the CEO's frustration. Can't OSHA or someone just hand us a standard definition of "world-class safety" we all can use?

But definitions, as we find with life in general, are not black and white. Go to most any dictionary and look up a word; you'll likely find multiple definitions.

The same holds true for another word — implementation. There is no one way to implement a world-class safety process. Oh, there is the popular Voluntary Protection Program template and a growing array of safety and health management systems. But all must be tweaked and refined by the participants "on the ground" to fit the templates and systems to the nature of the operation, its physical location and layout, the diversity of its workplace, and the values and beliefs of each organization's culture.

Leadership must confront this void created by having no authority figure available to tell you to do this, do that, and you'll get this — namely world-class safety results. Dr. Geller makes clear in this chapter — and throughout this book — that he is not addressing top-down leadership. To paraphrase the professor, he believes everyone in the organization has the opportunity every day to take the lead.

In Part 1 he emphasizes this, with one notable exception. People who do not like their job should not even try to be leaders. To put it in terms of People-Based Safety's™ ACTS, these disgruntled folks will not act like leaders, coach like leaders, think like leaders or see things as leaders should, according to Dr. Geller.

Also in Part 1, Dr. Geller offers three attributes he believes necessary for world-class leadership: frank talk about all safety incidents; setting behavioral expectations as soon as a new hire comes through the door; and the courage to confront both risks created by others and problems found in systems and in the physical environment.

Other characteristics of world-class leaders are described throughout this chapter. In Part 2, Dr. Geller uses the hedgehog concept put forth by Collins in his book, *Good to Great*, to make the point leaders must forge ahead with a constancy of purpose. They never veer or are distracted from their disciplined passion for safety.

Part 3 gets specific and gives you six good-to-great leadership qualities. Jim Collins uses them to describe great organizations. Dr. Geller is convinced they also define attributes of the best safety leaders.

Part 4 is a case study in leadership, based on the lessons learned from Hurricane Katrina and the floods that followed in August 2005. Many safety pros at some point will face a calamity — not on the scale of Katrina most likely — but calling for the leadership skills found in ACTS — quick actions; thoughtful counseling; concise, clear communication; quick thinking; and the ability to remove perceptual blinders and biases that inhibit effective problem solving.

Finally, Part 5 presents another case study in leadership, based on how leaders at Virginia Tech (where Dr. Geller has taught for 38 years) turned a horrible tragedy in which 32 students and faculty members were massacred, into a culture-enriching experience.

In their own way, Virginia Tech leaders exemplified PBS's ACTS. They took quick, specific *Actions* dealing with the media horde and the memorial ceremonies. As for *Coaching*, they placed a counselor at every classroom door when the students returned. *Thinking*... leaders did not "go negative," did not launch into immediate fault-finding, nor were they overly defensive in the face of Internet bloggers and reporters trying to find someone, something to blame. *Seeing*...with clarity, focus, and sensitive perceptiveness, university leaders recognized the healing that had to take place. A week to the day after the slaughter, Dr. Geller's Monday morning class was packed with 500 students. When the kids could have retreated, stayed home a little longer, they showed their commitment, determination and sense of unity. World-class leadership brings that out in people.

Dave Johnson, Editor
Industrial Safety & Hygiene News

1 World-Class Leadership: What Does It Take?

Recently, the CEO of a leading chemical company told me he dislikes the term "world-class safety" because it's so ambiguous. "Everyone talks about wanting a world-class safety program, but nobody provides a straightforward definition of this vision. What does it mean to be world-class?"

Well, let's take that ambiguity in a more personal direction: What does it mean to be a world-class leader?

I think we can learn lessons from organizations and programs and apply them to personal leadership attributes. For example, Jim Collins and his research team for *Good to Great*[1] spent five years studying 11 companies that rocketed from being good to achieving greatness. Collins' yardstick was financial — generating cumulative stock returns that on average were seven times better than general stock market — for a sustained period of 15 years.

Collins and his researchers systematically compared these good-to-great companies with a carefully selected set of 11 companies that maintained good profits for at least 15 years but never made the leap to true greatness, using Collins' financial criteria.

Good-to-great companies demonstrated signature qualities not consistently observed at comparison companies. These attributes afford us an operational definition of "world-class leadership" — and suggest ways for you to achieve this enviable level of leadership.

Here are Collins' five fundamentals:

1) Get the right people on the job;

2) Get the wrong people off the job;

3) Match talent and interest with job operations;

4) Maintain a climate of truth-telling by engaging people in rigorous debate, analysis, and continuous learning; and

5) Confront the facts, even when they are harsh.

The "right" stuff

Collins stresses from the start that an organization needs to employ the right people — "it's who you pay, not how you pay them," he writes. He uses a bus as a metaphor for today's organization, emphasizing the need "to get the right people on the bus in the first place and to keep them there."[2]

Character, work ethic, conscientiousness, and values are the "right" stuff to look for in the right people. These qualities are more important than educational background, practical skills, specialized knowledge and work experience, according to Collins. Skills and know-how are teachable; experience changes with time. But things like character and values are presumably more permanent traits.

Top people — leaders — are motivated by the intrinsic or natural consequences of their job, according to Collins. If people do not find such satisfaction in their job, it's in the best interest of all involved to let them go early or find them another assignment.

In fact, I contend everyone in an organization can find their seat on the bus and be a leader — with one exception. People who do not like their job, find no satisfaction in it, should not even try to be leaders. To put it in terms of People-Based Safety's™ ACTS, these folks will not *act* like leaders, *coach* like leaders, *think* like leaders or *see* things as leaders should.

Truth-telling

Many of you reading this book are not in the position to hire or fire people — push people off the bus, in Collins' term. You very well might not have the authority to yank people from one seat to another. Managers hire

WELL, IT'S ALL IN HOW YOU DEFINE "CHOP."

Leaders own up and tell the truth.

and fire and transfer people. I'm not talking here about management, but rather leadership. And let me be clear: management and leadership are two different things.[3]

At every level of an organization, leaders, no matter what their job, should engage in rigorous debate, analysis and continuous learning to uncover and report the objective facts of current reality as it relates to the scope of their task. This calls for leading with questions — not answers. And leaders seek facts — not faults.

Leaders, in the course of doing their job, do not shy from adversity. To paraphrase Collins, they confront the "brutal facts" of their situation (poor equipment, poor procedures, etc.) and confront them head-on. They emerge from their troubles stronger than before.

Relevance to leadership

As I mentioned, the special qualities of good-to-great companies can help us define world-class safety leaders. By the way, these qualities will also serve to enrich your organization's culture. Imagine a company with hundreds or thousands of individuals, leaders in their own domains, contributing these qualities every day.

1) World-class safety leadership requires open, frank, and fact-finding

conversations about all safety-related incidents, from close calls and first-aid cases to the most serious injuries and fatalities.

World-class safety leaders discuss freely and openly, without embarrassment, all injuries (minor and major), as well as close calls. Leaders realize that only through such open discussion can the environmental, behavioral, and cultural factors contributing to these mishaps be rooted out. Facing such adversity head-on results in leadership more empowered to prevent occupational injuries.

2) The standards of world-class leadership are planted early on; when individuals are first hired, they receive specific safety-related behavioral expectations.

3) When you are working with someone whose safety-related behavior does not meet specific expectations, a corrective action plan is implemented. Now, it very well may not be your responsibility to develop the plan and be the person to engage in candid conversation and elicit a personal commitment to change — or to get the poor performer off the bus. But safety leaders, simply in the course of doing their everyday assignment, cannot overlook risks or hazards created by the at-risk behavior of others.

At a workshop I once gave to 90 first-line supervisors, the plant manager urged me to emphasize this particular lesson. In some cases, an employee or contractor may not care about safety to the degree demanded by the work culture. He wanted his supervisors to look for these incongruities and then engage in open and frank conversation with individuals whose at-risk behaviors suggest they are not prepared or appropriate for a particular job.

I extend this act of *seeing* (recall ACTS) beyond supervisors. Everyone has the responsibility to be alert to unsafe acts. And everyone can offer corrective feedback to a coworker as part of formal or informal coaching. (I address this in more detail in Chapter Five, Improving Communication.)

To be a world-class leader in safety, you can't duck the brutal reality that at-risk behavior cannot exist in a workplace that promotes safety as a core value. This does not necessarily mean a risk-taking person should be bumped off the bus, but it does mean some corrective action is required — and the sooner, the better.

Part of that corrective plan requires the person to commit to specific behavior change, and any peer or management support needed to make this happen must be detailed.

As I mentioned, we are talking here about the qualities of safety leaders and an ideal Total Safety Culture. These qualities reflect safety ideals to which we should aspire.

2 Are You a Fox or a Hedgehog?

As I mentioned in the previous section, engaging in open and frank assessment and conversation regarding safety problems, risks and hazards is one of the qualities of a safety leader. Again, you don't need to be the plant manager, a supervisor, or even the safety and health manager, to be a leader and be open and frank about safety issues.

According to Jim Collins in his book, *Good to Great*, "One of the primary ways to de-motivate people is to ignore the brutal facts of reality."[4] Regarding safety, this means leaders at all levels of the organization must do their part to ensure co-workers adjust their behaviors to be consistent with those required for a Total Safety Culture.

Winning is more than looking good.

The Hedgehog Concept

In order to typify the simple and organized focus of the good-to-great companies, Collins contrasts the hedgehog with a fox. We can apply the same concepts to safety leadership.

The fox looks like a winner: crafty, quick, and fleet of foot.

The hedgehog waddles along, day after day, focusing on the bare necessities of living.

Analogously, some people are showy, diffuse, and scattered with regard to purpose, goals, and action plans. Others are more like the hedgehog, simplifying their complex world into a single unifying principle or vision that provides organization and focus for their daily activities.

Good-to-great companies (and leaders) did the latter, which Collins and his research team label "The Hedgehog Concept." In terms of leadership, leaders have profound answers to these three questions in order to organize and focus all their activities — no matter how narrow or broad their job might be:

What can I be the best at?

What is the payoff for my safety leadership?

What am I deeply passionate about?

This Hedgehog Concept reminds me of the "constancy of purpose" principle advocated by W. Edwards Deming.[5] Leaders discriminate between a) what they can do best and what they cannot, b) what is profitable for them and what is not, and c) what they are passionate about and what they are not. These discriminations define their mission, and fuel their goal-setting and action plans.

Relevance to safety

World-class safety leaders (and cultures) are hedgehogs when it comes to injury prevention. They understand what it takes to be among the best in industrial safety, and believe they can reach this level of excellence. They also realize the direct correlation between their financial profit and their success at preventing injuries. Their passion to be world-class in safety alerts them to any inconsistencies between this vision and various company activities, from strategic planning in the boardroom to behaviors on the shop floor.

Disciplined leadership

Maintaining hedgehog principles and constancy of purpose requires discipline. But the meaning of discipline here is not punishment, as this term is often used in industry. I am referring to self-discipline. Leaders do not require, nor even want, top-down controlling discipline from others.

"Sustained great results depend upon building a culture full of self-disciplined people who take disciplined action, fanatically consistent with... the Hedgehog Concept."[6]

The Hedgehog Concept in operational terms defines a consistent system with a clear mission as well as constraints. Within this tight structure, leaders have opportunities to choose a particular course of action. But leaders must be responsible and self-disciplined.

The practice of discipline

Collins delineates certain practices of the good-to-great companies that help define a culture of discipline. These suggest guidelines for becoming a leader in your own work domain.

Foremost is strict adherence to the Hedgehog Concept. Leaders are focused individuals. For leaders, a "stop doing" list is as important as a "to do" list. Leaders continually ask: "What am I doing to be the best? What am I doing that detracts from me being the best?"

In terms of your organization's culture, this sort of discipline in safety means dropping policies, programs, and slogans that do not contribute to safety excellence. For example, I've met several safety pros who realize their organization's safety incentives or bonus system rewards a reactive mindset and detracts from proactive involvement in injury prevention. Their challenge is to gain the support needed to drop or alter this ineffective program. (See Chapter Five, Improving Communication for more on how to ask for support.)

In addition, while special safety efforts often start off with lots of participation and optimism, many fizzle over time. Obviously, such drifting beyond original intentions detracts from optimal performance. If such programs cannot be re-energized or refined to get back on track, they should be dropped. (Chapter Four, Enriching Your Culture, offers ideas for powering participation).

Leaders do not operate out of a fear of failure but focus on achieving success. And they recognize there are no magic bullets or quick fixes (all too common thinking in safety). Rather, success evolves from a series of incremental changes or small wins.

In similar fashion, a Total Safety Culture promotes success-seeking over failure avoiding by putting more focus on the daily proactive things people do to prevent injuries than on the injuries themselves. These cultures define their safety excellence by the various safety-related activities accomplished each day to prevent mishaps, not by such reactive, failure-focused outcome statistics as total recordable injury rate and workers' compensation costs.[7]

**Discipline and focus are key
to success.**

To conclude

While Collins presents the Hedgehog Concept as a company guideline to achieving greatness, I hope this presentation shows you I find this theory equally relevant to individual leadership. For example, happy, self-motivated leaders perceive they are well paid for applying their special talents effectively on a job they feel passionate about doing well.

People who do not believe they are applying their talents effectively for important work are not self-accountable nor intrinsically motivated. They are anything but leaders. They cannot be the best they can be — and can detract from the achievement of "world class" in safety. Effective leaders can sometimes help these individuals reframe their thinking and develop a relevant Hedgehog perspective.

3 Defining Good-To-Great Leadership

The six qualities listed below distinguished the leadership of the good-to-great organizations from the leadership of the comparison companies, as described in Jim Collins' book, *Good to Great*.[7] I'm convinced they define attributes of the best safety leaders.

1 — Personal humility

"Good-to-great leaders never wanted to become larger-than-life heroes," rather they "were seemingly ordinary people quietly producing extraordinary results".[8]

2 — Acknowledge contributions

Good-to-great leaders attribute company success to factors other than themselves. As systems thinkers, they see the big picture and realize their success is contingent on the daily small-win accomplishments of many individuals. Thus, they acknowledge the synergistic contributions of many others who enable remarkable results.

3 — Accept responsibility for failure

Good-to-great leaders face the brutal facts of less-than-desired outcomes, and hold themselves accountable without blaming other people or just "bad luck." Leaders of the comparison companies too often blame others for lack-luster performance while taking personal credit for extra-ordinary results. Social psychologists call this the "self-serving bias."[9]

Some people blame others for their own mistakes.

4 — Promote a learning culture

Humble leaders are always learning, with impassioned belief in never-ending improvement. They lead with questions rather than answers, and promote frank and open dialogue and debate. The result: People are not satisfied with the status quo, but are engaged in finding ways to improve company performance.

5 — Work to achieve, not to avoid failure

Good-to-great leaders never waver in their resolve for greatness. Failure is not an option; it is not even considered. With an optimistic stance, these leaders focus on achieving exemplary success.

At the same time, they adhere fervently to the Hedgehog Concept, as I discussed in the previous essay. This means identifying: a) what you can do best, b) what you feel passionate about, and c) what is profitable. Then attend to your envisioned enterprise with fanatical consistency and disciplined purpose.

6 — Encourage self-motivation

Self-motivation is key to long-term productivity and is gained through intrinsic consequences. In other words, people are self-motivated when their behaviors provide natural ongoing consequences that are rewarding.

When does behavior on the job become intrinsically reinforcing and self-motivating?

Answer: When people believe their work is meaningful.

When does this happen?

Answer: Sometimes the special value of the work is obvious, as when people are engaged in activities that prevent injuries. But even in these cases, it's critical to give the kind of interpersonal attention that reassures people they are accomplishing meaningful work. Great leaders know how to do this, and do it often.

I think this final quality is most significant for safety, because it defines the source of the motivation that keeps effective safety leaders going. Working for safety is meaningful work that fuels self-motivation.

Jim Collins ends his book with the following: "It is impossible to have a great life unless it is a meaningful life. And it is very difficult to have a meaningful life without meaningful work."[10]

People-based safety leaders do meaningful work — and have meaningful lives.

4 Hurricane Katrina Leadership Lessons

Hurricane Katrina and the flooding of New Orleans and parts of the Gulf Coast captured the nation's attention in August 2005. Sadly, better leadership at the federal, state and local level could have prevented so much of the destructive aftermath from that hurricane.

What did we learn from this catastrophic event? Did our direct or vicarious connections with the devastation and human tragedy teach us anything?

Here are six lessons of leadership related to both Hurricane Katrina and industrial health and safety.

Lesson 1 — Leaders accept that accidents happen

For more than two decades I've advocated that safety leaders be cautious in their use of the word "accident" — a term that implies chance or lack of personal control. Most injuries are caused by manageable factors, and leaders need to know this, but some do involve unknown or uncontrollable determinants. Hurricanes exemplify the latter. Leadership can prepare to avoid the negative consequence of a hurricane with protective devices or escape plans, but leaders can't stop or redirect a hurricane.

Similarly, it's important for safety leaders to own up to the fact we don't know enough yet to prevent all unintentional injuries. Accidents still happen in the workplace, in the home, and on the road. Like hurricanes, these are the unpreventable mishaps that require protective barriers and emergency escape plans. When leaders believe unpreventable and dangerous incidents are indeed possible, they are more likely to take the lead in emergency planning.

Lesson 2 — Leaders envision and communicate tragic consequences

I've heard many safety leaders say, "What we need around here is a fatality to get management support for safety."

Of course, no one really wants such misfortune. But People-Based Safety™ leaders can use a workplace tragedy to focus people's attention on the actions needed to prevent a recurrence. Electronic images of environmental destruction and human suffering after Hurricane Katrina motivated quality responsiveness to hurricanes Rita and Wilma by local, state, and federal agencies.

But how long will the motivating images of Hurricane Katrina's wrath last? For those not directly impacted by the hurricanes of 2005, these images will likely fade with time, along with the proactive and protective behaviors motivated by such visualization. People will eventually forget the devastating consequences of failing to prepare for an infrequent but inevitable accident.

At work, a preparedness mindset triggered by tragedy ultimately returns to "business as usual," as well as defensive denial supported by self-talk like "It can't happen again." But ask NASA officials who suffered through the *Challenger* and *Columbia* space shuttle explosions. It can happen again.

Fortunately, powerful images of injuries or near-injury, and the associated emotions, can be revived by safety leaders. The relevant safety behaviors are revisited and activated by leaders through storytelling techniques.

Lesson 3 — Leaders invest in people

"What is the return-on-investment (ROI)? And how long will it take to see results?"

These are the two questions asked most frequently by managers considering implementation of a particular safety program. They want to know how much the process will cost and how long it will take before financial advantages are realized.

We're told it would have cost about $2.5 billion to renovate the Lake Pontchartrain levees so they could protect New Orleans from flooding after a Category 4 or 5 hurricane. That cost is certainly mind-boggling, which is probably why the levees were not upgraded.

The chance a Category 4 hurricane would hit New Orleans seemed remote, so the city reaped soon, certain, and positive consequences by investing in other structures or business ventures. Because safety was not a priority (let alone a value), the cost of rebuilding New Orleans will amount to more than $200 billion. These financial estimates result in an overwhelming cost-benefit ratio in favor of investing in levee improvement. Of course, hindsight is 20/20.

Although the financial cost of Hurricane Katrina's destruction was astronomical, the total cost cannot be tallied in dollars. We can't put a price tag on the human pain, suffering, and deaths caused by the flooding of New Orleans. We saw horrific scenes of people stranded on streets, on rooftops, and in the Superdome. We heard numerous reports of people drowning because they couldn't escape the floodwaters. Emergency crews and life supports were unavailable, delayed, or misdirected.

Bottom line: The aversive consequences from the lack of proactive protection and emergency planning for Hurricane Katrina put a critical spin on the investment term "ROI." Intervention for safety is an investment in people, and safety leaders know the ROI is far more than financial savings or business profits.

Lesson 4 — Leaders know frustration and pain elicit aggression

The aftermath of Hurricane Katrina included substantial aggression. Electronic images on the Internet and national television exposed the public to outspoken anger, rampant looting, and interpersonal assaults among those trapped in New Orleans. Were these people criminals or hoodlums with no regard for others?

Research with many varieties of animals, including humans, has shown that frustration and pain cause aggressive behavior. Much of the outrage and hurtful encounters observed after Hurricane Katrina were due to circumstances, not character flaws.

This lesson is relevant for safety leaders operating in everyday work situations.

Obviously, the workplace does not harbor the amount and severity of pain and frustration as that linked to Hurricane Katrina. But as the result of emotional upheavals, frustrations, or disappointments at home or at work, individual workers can be in psychological states approximating those of the hurricane victims. I suggest you, as a people-based leader, raise this possibility in one-on-one conversations or group meetings, and discuss workplace situations that cause employees to feel frustrated. By identifying and removing frustrating conditions, leaders decrease anger and interpersonal conflict conducive to risky behavior and unintentional injury.

Lesson 5 — Leaders own up and apologize

In this chapter, I have discussed qualities of leaders defined by Jim Collins in his research-based book, *Good to Great*.[11] Humbly admitting mistakes and apologizing headed this list. These virtues were not displayed by local, state, and federal leadership queried about Katrina effects. For example, the mayor of New Orleans blamed the Federal Emergency Management Administration (FEMA), while the head of FEMA passed the buck to the mayor and the governor of Louisiana.

Two weeks after Hurricane Katrina, the White House claimed responsibility for the lack of timely and competent federal responsiveness to the aftermath of this catastrophe. But I heard no explicit apology nor saw emotional empathy for the disastrous consequences of delayed and incompetent attention to public safety and health. Did I miss something?

I've heard safety consultants assert effective leaders are "emotionally resilient" and exhibit little emotion. Although many in charge of helping the victims of Hurricane Katrina seemed to fit this description, I know of no research supporting this prescription. Rather, my experience suggests the opposite. I'm convinced the best leaders feel true empathy for victims and thus cannot help but show some emotion. And when people see emotional reaction from their leaders, they believe their leaders are empathic and genuinely care.

Leaders admit personal mistakes and move on.

Lesson 6 — Leaders actively care

Media coverage of Hurricane Katrina's effects was not completely negative. Remember the varied and extensive helping hands reaching out to hurricane victims. Many traveled hundreds of miles to actively care, and many more sent financial support. In their own way, each of these people took the lead and acted with care, empathy and responsibility.

In the workplace, I suggest you as a people-based leader can collect personal stories and relive testimonies of people going beyond the call of duty to assist others. People will actively care when given appropriate direction. When this direction is given by modest leaders who own up to their own human weakness and make an emotional appeal for help, even more will actively care.

Your role

People-Based Safety™ leaders should help ensure the public never forgets the calamitous destruction and human misery that occurred because appropriate preventive action was not taken in New Orleans. With billions of dollars, New Orleans may be rebuilt and protected with the proper levee system, but the money can't bring back the hundreds of lives lost nor restore the historic structures, memorable possessions, and priceless keepsakes washed away by floodwaters.

I hope we never let people forget that most of these losses could have been prevented with the types of proactive plans and activities advocated daily by safety leaders. And may we find consolation and encouragement in observing that many people will react to tragedy by actively caring, as many did for those victimized by the hurricanes of 2005.

5 Leadership After Overwhelming Tragedy

In their book, *Measure of a Leader,*[12] Aubrey and James Daniels claim the best way to determine the quality of leadership is to evaluate the behavior of the followers. I discuss this measurement concept further in Chapter Three, Part 4. Here I want to underscore that point by describing the unity and fierce loyalty I witnessed by the Virginia Tech (VT) community in the wake of horrific tragedy.

I also want to tell you of some of the numerous examples of exceptional leadership I saw. But keep in mind, as Virginia Tech President Charles Steger said the day following this tragedy, "Words are very weak symbols of our emotions."

"A gunman is loose"

I am writing this just a week after the most devastating day of my life. As you now know, on April 16, 2007, 27 VT students and five professors were gunned down by a 23-year-old senior student. That morning, I was driving to the campus to teach 600 students in my Introductory Psychology class. My cell phone rang, and the coordinator of our Center for Applied Behavior

Systems told me to "Go home, our campus is locked down. We've been instructed to stay away from windows because a gunman is loose on campus."

Shocked and not quite believing what I had just been told, I returned to my home office. All day Monday I stayed glued to MSNBC as disturbing thoughts and emotions streamed though me: vexation, confusion, anger, sadness, grief, denial, and anguish. These cognitive and emotional states would intermittently invade my days and nights for weeks. And these emotions intensified when the media began reporting the personal stories of the victims, including acts of heroism by some killed or injured.

Coming together

On Tuesday afternoon I sat in the middle of our football stadium, in the midst of 10,000 students, faculty, and staff, watching a special convocation service on the large scoreboard screen. The event at nearby Cassell Coliseum, home of our basketball team, had quickly filled the arena's 9,000 seats and I was part of the outdoor overflow. This is when I began to perceive some positive aspects of this unthinkable massacre. The campus community — the Hokie Nation — was coming together to comfort each other. The University motto "Ut Prosim" — That I may serve — was coming to life.

(By the way, the Hokie Bird has been the official sport mascot of Virginia Tech since 1961. But in fact, a Hokie isn't anything at all, and has nothing to do with that orange and maroon mascot. It's a made-up word that was added to an 1896 football cheer by a student named O.M. Stull simply because it sounded good.)

At the service, leaders took turns trying to comfort us. After our university president clarified and validated the difficulty we all experienced in verbalizing the extreme emotions we felt, the governor of Virginia spoke with passion about belongingness and the need to "not lose touch with that sense of community." Then, President Bush asserted "schools should be places of safety and sanctuary and learning." He encouraged us to reach out to one another

The candlelight vigil at VT was memorable.

with the affirmation, "You have a compassionate and resilient community here at Virginia Tech."

That resilience was evidenced after the world-renowned poet and VT professor Nikki Giovanni delivered her rousing, almost defiant, oration.

We Are Virginia Tech
By Nikki Giovanni

We are Virginia Tech.
We are sad today, and we will be sad for quite a while. We are not
    moving on, we are embracing our mourning.
We are Virginia Tech.
We are strong enough to stand tall tearlessly, we are brave enough to
    bend to cry, and we are sad enough to know that we must laugh
    again.
We are Virginia Tech.
We do not understand this tragedy. We know we did nothing to deserve
    it, but neither does a child in Africa dying of AIDS, neither do the
    invisible children walking the night away to avoid being captured
    by the rogue army, neither does the baby elephant watching his
    community being devastated for ivory, neither does the Mexican
    child looking for fresh water, neither does the Appalachian infant
    killed in the middle of the night in his crib in the home his father
    built with his own hands being run over by a boulder because the
    land was destabilized. No one deserves a tragedy.
We are Virginia Tech.
The Hokie Nation embraces our own and reaches out with open heart
    and hands to those who offer their hearts and minds. We are
    strong, and brave, and innocent, and unafraid. We are better than
    we think and not quite what we want to be. We are alive to the
    imaginations and the possibilities. We will continue to invent the
    future through our blood and tears and through all our sadness.
We are the Hokies.
We will prevail.
We will prevail.
We will prevail.
We are Virginia Tech.

At the conclusion of Professor Giovanni's short speech, students inside and outside the coliseum clapped, cheered, and chanted "Let's go, Hokies!" Perhaps some emotional healing had already begun.

Because classes were canceled for the rest of the week, I wondered how many students would attend the eight o'clock vigil scheduled for Tuesday evening. The vigil was planned and organized by Hokies United — a student-run alliance of several VT organizations.

I was amazed. Thousands of students, faculty, and community residents gathered on the 40-acre Drillfield at the center of the VT campus. Each of us received a candle donated by local businesses, resulting in a sea of small "points of light."

Dr. Zenobia Hikes, the VT vice president of student affairs, gave the only speech. "We will move on from this, but it will take the strength of each other to do that. . . . We are a community of strength. We are a community of pride. . . . We are a community of compassion," she declared.

Buglers then played "Taps." The crowd stood silent for many minutes, holding their candles high. After about 30 minutes, a group of students cheered, "Let's go. . ." and a group in the opposing half of the field yelled, ". . . Hokies!" The chanting got louder and louder until "Let's go, Hokies!" could seemingly be heard for miles.

We were not the only campus to hold a vigil on Tuesday evening. Throughout Virginia and beyond, churches, colleges, and neighborhoods held candlelight vigils to grieve for the victims at VT. Many ceremonies included bells or chimes sounding 32 times — once for each of the fallen Hokies. Nationwide, people showed actively-caring sympathy for the pain our VT community was feeling. People near and far were wearing the orange and maroon VT colors. Our Atlantic Coast Conference sports rivalries melted away as universities expressed their condolences. A bridge at the University of Virginia was painted "Hoos for Hokies."

As the days went by, blossoming trees on campus were draped with black, maroon, and orange strips of cloth tied around their trunks. Some students and faculty stood in circles at the center of the Drillfield, hugging, praying, and singing hymns. Others sat on the grass to watch the scene and ponder the horrible fate of colleagues and classmates.

An astounding remark

Thirty-three Hokie stones were placed in a semicircle around the podium at the head of the Drillfield, each stone topped with flowers, an American flag, and a Virginia Tech pennant. Each stone included the name of a victim, and notes and memorabilia commemorating the life of the individual represented.

Yes, there was a memorial stone for the killer. But by Friday, the Hokie stone for the gunman was gone, though the flowers and his name card remained. Apparently, some mourners could not accept this commemoration for the gunman. Who could blame them?

I told the associate dean for the College of Science, a clinical psychologist and previous head of our Department of Psychology, I was dedicating this book to the 32 fallen Hokies.

"We lost 33 Hokies on Monday," he responded.

I was surprised and astounded by his remark. "Can you forgive this killer?" I asked.

"It's not about forgiveness," he said, "but about recognizing this individual was mentally ill and his family grieves for their loss."

I walked away thinking, "Our University is so fortunate to have this individual in a key leadership position."

Acts of caring

During the week classes were canceled, the VT leadership disseminated numerous thoughtful emails, addressing ways to aid the healing process, and detailing protocol for handling classes and assigning grades. All these leadership decisions were specified as "student-centered." And the faculty was urged to be student-centered in all decisions involving students.

Our campus was bombarded by actively-caring voicemail, email, cards, and poster displays from other universities throughout the week. A "condolence link" was established on the VT website, and before the week ended, more than 25,000 entries were logged, covering 81 pages.

I personally received more than 200 emails communicating concern and compassion for our plight. In addition to past students, colleagues, and current acquaintances, people whom I had never met personally, or met only once many years ago expressed sincere condolences.

From a Mountaineer to a Hokie

Excerpts from an email...

We Mountaineers are with you, Virginia Tech....

We mourn with your administration, the members of which will spend the rest of their lives with unnecessary guilt, even though they did all they could do.

We mourn with your campus police, heroes who put their lives on the line.

We mourn with the parents.... We even mourn for the mother of Cho-Seung Hui, who will spend the rest of her life knowing that her son pained a nation and will forever wonder what she did wrong.

Virginia Tech, we are mourning with you — but we are also celebrating, for the heart and soul we have seen in Blacksburg gives us hope that there is goodness in the world, after all.

Do you realize what strength you have demonstrated, Virginia Tech?

Do you realize how beautifully you have banded together?

Do you realize what courage you have shown?

Do you realize the inspiration you have provided?

Hokies, in the face of such horror and such emotion, you should be proud of the way you handled yourselves this week. Your brothers in Morgantown applaud you for your strength, and we stand behind you in your recovery.

Many emails to our entire University community were inspirational. The example on page 44 is from a university-wide email from West Virginia University — our sports arch-rival for many years.

Invasion of fault-finders

The media siege on campus was not especially pleasant. "Hokie Nation Needs to Heal — Media Stay Away" proclaimed a large neon orange sign. Throughout the first week, TV anchors — Katie Couric from CBS, Brian Williams from NBC, Greta Van Susteren from Fox, and Larry King from CNN — reported from Blacksburg, and the clear focus of most media coverage was identifying causes of the tragedy.

Why was the campus not locked down during the two-hour delay between the killings in the dorm and the classrooms?

How was a student with such a demeanor able to reach his senior year at VT?

Why was the killer treated as an outpatient from the local mental health facility?

How could a person found at risk for hurting himself or others purchase two handguns?

How could someone walk across the center of campus in broad daylight with guns, several rounds of ammunition, and chains to lock a classroom building from the inside?

News reporters filled the air with fault-finding questions, which VT's leaders deftly deflected. Friends and family of victims were confronted with such loaded questions as "Are you going to return to Virginia Tech after this disastrous event?" "Are you angry with the university for their inadequacy in preventing this incident?"

The most despicable stunt was showing the videos the gunman had sent NBC. Instead of turning the horrid scenes over to the FBI and merely describing the content in a news report, NBC followed the killer's wishes and made him infamous. This was a clear lack of leadership and emotional intelligence.

Despite the "witch hunts" and VT slamming, student reactions were overwhelmingly positive. "Of course, I'm returning to Virginia Tech. I love this place." "We don't blame anyone but the shooter for this terrible happening." "We Hokies stick together, and we will survive."

Measure of leadership

What does all of this have to do with leadership? I say "everything."

The Hokie spirit, inspired by leaders with clarity, focus, and insight, is encapsulated by the following excerpts from a campus-wide email from our Dean of the College of Science: "Virginia Tech is still a vibrant and nurturing community. . . . We are bruised, but we are not daunted. Even after the reporters leave, and national and international attention turns elsewhere, we will still be here for one another, and we will remember."

Classes resumed April 23, one week after the slaughter. University leaders were organized and well-prepared. More than 250 mental health counselors, including several graduates of our Ph.D. program in clinical psychology, traveled long distances to be available for our students. Every class in which the victims were enrolled had at least one counselor and staff volunteer on hand throughout the class. Three counselors were assigned to each of my large introductory psychology classes.

But was all the preparedness necessary? Would VT classes be well attended? Yes indeed. We were all surprised to see so many students in our classes. Practically everyone returned and attended their Monday classes — another measure of campus leadership.

Spontaneous leadership

Throughout our ordeal, which is far from over, we witnessed numerous examples of leadership from every dimension of our campus community. And one leader's actively caring efforts enabled helpful leadership from others. Here's an example:

On Monday afternoon, April 16, Tod Whitehurst, a VT employee and a nationally certified massage therapist (CMT), was sent home from the devastated campus. He immediately got on the phone to local members of the massage community and to the Blue Ridge School of Massage & Yoga, where he is a part-time instructor. Within 24 hours, he and Valerie Beasley, CMT, also a graduate of the school and a member of the local Red Cross, put massage therapists to work in churches, at the Inn at Virginia Tech where grieving families were gathering, at the University's Cook Counseling Center, and other Virginia Tech locations. Both Tod and Valerie spent long hours on campus, providing nurture and stress relief to students, families, staff, faculty, EMTs, police, counselors, and clergy. On-site massage continued at various locations on and off campus through May 10.

Bottom line: It takes world-class leadership to bring the best out in people in such trying circumstances as we experienced at Virginia Tech. I am extremely proud to be a 38-year veteran of our Hokie Nation, and I am eternally grateful for the special leadership that enables and empowers us all to be the best we can be. We will prevail. We are Virginia Tech.

CHAPTER THREE

Developing Your Leadership Qualities

Introduction

"How do we get to the next level, Dr. Geller?" In this chapter, the good professor reiterates a crucial theme of *Leading People-Based Safety*™ — everyone is in the position to develop personal leadership qualities.

As he says in Part 1, "For safety performance to improve beyond current plateaus, hourly workers need to provide more advice, involvement, and interpersonal accountability. In other words, we need hourly workers... to speak up."

To speak up, we all need to find our voice, the topic of Part 1. Dr. Geller shows you how you can take your *talent*, your *passion*, what you have to *contribute*, and your guiding *conscience* (Stephen Covey's four characteristics of "voice") and use them to help take safety to the next plateau, and enrich your culture in the process.

Find your voice, and you are strengthened to lead, in whatever job you have in your organization. As described in Part 2, you will find opportunities to:

• *love* (form non-romantic, productive relationships);

• put your *energy* to use;

• be *audacious* and challenge your organization to go beyond old traditions (such as safety's 3Es of engineering, education and enforcement) to reach higher levels of performance; and

• use *data* (evidence, documentation, records) to back up your creative, innovative safety ideas that go beyond the traditions rooted in your culture.

Love, Energy, Audacity, Data — together they form the acronym LEAD — defining qualities of safety leaders.

Part 3 focuses on another quality of leaders: the ability to be the real deal, to be credible, respected, trustworthy, and authentic.

How do you "walk authentic talk"?

Dr. Geller points to numerous leadership behaviors, from being open to feedback, vulnerable, and exerting self-discipline, to helping workers become self-directed, giving responsibility and recognizing a job well-done, treating adults as adults, and simply asking "How can I help?" improve safety around here.

Data, the "D" in Dr. Geller's LEAD model, might confuse or seem out of place to some safety pros. In Part 4, Dr. Geller draws the connection between data (injury records or audit findings, perhaps) and accountability. Data directs and motivates behavior, and by extension, accountability. Leaders hold people accountable for data — safety performance indicators — they can directly influence.

Speaking of keeping score — a favorite pastime in safety — many safety leaders strive for a score of zero incidents. In Part 5, Dr. Geller shows how setting the bar high can set up leaders for a failure-avoidance mindset. Avoiding failure at all costs is not a characteristic of authentic, inspiring leaders. Safety leaders do not accept failure. They don't go for "quick fixes" and "magic bullets" to circumvent failure. And they don't become complacent when they achieve success — zero incidents or whatever the vision may be. As mentioned earlier, leaders maintain a constancy of purpose and inspire others to "buy into" that mission, to take safety performance to the next level, and enrich their culture.

Dave Johnson, Editor
Industrial Safety & Hygiene News

1 Finding Your Voice

Author Stephen Covey refers to the Industrial Age as the "Age of Control."[1] Manual workers are expected to follow the rules and standards defined by top management. Workers do physical labor, but leave their minds at the door. They become institutionalized to speak to their supervisors only when spoken to. They essentially lose their "voice."

Global competition today requires people to work smarter. This is the Age of the Knowledge Worker. To work smarter and compete successfully, corporate decision-makers need more input from their employees. For safety performance to improve beyond current plateaus, hourly workers need to provide more advice, involvement, and interpersonal accountability. In other words, we need hourly workers — once silenced by the top-down control paradigm — to become knowledge workers and speak up. They need to find their "voice," and use it to benefit their workplace.

Certain leadership qualities enable employees to offer safety-related advice and increase engagement in safety processes. They are captured in the acronym LEAD, which I explain later in this chapter, to classify key qualities of effective people-based leaders.

Before I discuss the specific leadership qualities reflected in LEAD, I want to first emphasize the importance of finding your "voice" as a leader.

Finding your voice

Over the years, university students have frequently asked me, "How should I decide what major course of study to pursue?" Then during their senior year many ask, "How can I know what career goals to strive for?"

My first answer to their questions has consistently been, "Find a fit between function and talent, and find your voice." I use "find your voice" as a metaphor to express the need to seek compatibility between tasks (as in a job assignment) and personal interests, natural talents, and learned skills.

Sometimes I relate the personal story of when I turned down my candidacy to become head of the Virginia Tech Department of Psychology in 1980. I was considering the advantages of such a promotion, including a large salary increase and greater control over departmental business, when an esteemed colleague entered my office and asserted, "Don't do it. You are a researcher and a teacher, not an administrator."

I followed the profound advice of my friend and colleague who had served as head of the Psychology Department at Carnegie Mellon for 15 years. He knew my "voice" better than I did at the time.

Today I know my "voice" better, and am so grateful I listened to my wise colleague. Through much reflection on life experiences, I've increased my appreciation for this "find your voice" metaphor. Stephen Covey's book, *The*

8th Habit: From effectiveness to greatness[2], has added substantially to this realization.

The 8th Habit

Most readers are familiar with Dr. Covey's first bestseller, *The 7 Habits of Highly Effective People.*[3] Indeed, several safety consultants have incorporated the seven "habits" into their training and consultations. A few, including myself, have given keynote addresses and professional development workshops with titles like "The 7 Habits of Highly Effective Safety Leaders" (*See sidebar.*)

The 8th habit proposed by Dr. Covey is "Find your voice and inspire others to find theirs".[4] According to Covey, helping people find their voice is a primary mission of leadership. The leader who convinced me 27 years ago to withdraw my name from department-head consideration not only convinced me my voice was not in administration, he also called special attention to my accomplishments that implied talents and skills for teaching, research, and scholarship.

Talent vs. skill

Is there a difference between talent and skill? The *American Heritage Dictionary* (1985) does not make a clear distinction, defining talent as "a mental or physical aptitude; natural or acquired ability"[5] and skill as "proficiency, ability, or dexterity."[6]

However, Covey maintains talent to be one's natural gifts and strengths, whereas skills are learned or acquired. Talents require skills, but people can apply their knowledge and skills in areas where their talents do not fit. Then people go through the motions with competence, but it takes external accountability systems to keep them going.

I have a hard-working colleague, for example, who has superb knowledge and research skills relevant to each university course he teaches. However, he is not a talented teacher. He prepares well for each class but does not connect theory and concepts to student-relevant examples, and he does not activate critical thinking nor engage the students in interactive discussion. Clearly, teaching is not this individual's "voice;" rather his "voice" is in research and administration.

Four aspects of "voice"

Covey claims a person's voice "lies at the nexus of:
   • *talent* (your natural gifts and strengths),
   • *passion* (those things that naturally energize, excite, motivate, and inspire you),
   • *need* (including what the world needs enough to pay you for), and
   • *conscience* (that still, small voice within that assures you of what is right and that prompts you to actually do it)."[7]

Dr. Covey links the italicized words in the preceding quote to the four dimensions of the whole person — mind (for talent), heart (for passion), body (for needs), and spirit (for conscience). Let's explore these four aspects of people as they relate to the life of People-Based Safety™ leaders.

A special "voice" for safety

The challenges of safety promotion and injury prevention require unique skill, *talent*, and *passion*. But the foundation of this special calling is *service*. However, such service is not always appreciated by those served, and the financial remuneration to cover your efforts is rarely equitable. The typical safety professional encounters many employees who earn more money than they do, yet work less diligently and put in fewer hours.

In addition, safety professionals accomplish their daily tasks with less-than-adequate resources. Often their requests for materials and/or personnel support are seemingly ignored. Many don't dare ask for financial aid and time off to attend a professional development conference.

Why is safety and those who work for injury prevention treated this way?

Because the benefits of safety efforts are not readily visible. Regardless of the degree of attention to safety, injuries are relatively rare. And often it takes a serious injury or fatality to get corporate leadership to evaluate their safety policies, programs, and processes.

So what keeps the safety professional going under these trying circumstances? Where is their "voice"? Clearly, the challenge of safety leadership — in the management sense — is not for everyone. Besides certain knowledge, skills, and talents, safety leadership requires a high level of emotional intelligence — the ability to delay immediate gratification for delayed and potentially greater rewards.

And what is their reward? You know — the prevention of an injury or fatality. But how do we observe this benefit of safety and reap proper recognition and appreciation? That's the problem, and the reason safety professionals are among the unsung and under-compensated heroes of our society.

Thus, very special people pursue the challenge of safety leadership — people who find their "voice" in serving others, but don't need soon, certain, and positive consequences to affirm their value. Their *passion* to help others is supported by *spiritual* intelligence — that still, small voice within that transforms passion into compassion and says, "I might be sacrificing but I know I'm doing right because I'm serving others." If you find your "voice" here, you are a special person.

I want to end by emphasizing one important point: while I say the challenge of safety leadership in the management sense is not for everyone, a culture rich in safety values needs the passion and talents of every individual in the organization. Everyone can apply PBS's ACTS — Acting,

Coaching, Thinking and Seeing — to enrich their work culture and help keep people safe.

Habits of Effective Safety Pros

Be proactive
To build safe behaviors, define a target behavior and a rewarding or correcting feedback consequence.

Set goals
"Begin with the end in mind," as Covey says — tell employees what the rewards or unpleasant consequences will be when a specific goal is reached.

Know your priorities
"Put first things first," says Covey — keep aware of your shifting personal priorities and determine what really needs to be done to affirm your values.

Make safety win-win
Set up a team safety goal, a contract between members of the team, with all team members agreeing to the benefits of the goal and the behaviors required to achieve them.

Teach listening skills
"Seek first to understand," says Covey — a powerful, genuine way to show employees you actively care.

Build synergy
Bring employees, or a team, together to collaborate on definitions of safe and at-risk conditions and behaviors; audit or otherwise observe those behaviors and conditions; and step in when necessary to support safe behaviors and correct at-risk behaviors and conditions.

Strive to improve
Covey calls this "sharpening the saw" — to avoid complacency, keep defining and achieving small-win safety goals that reflect continuous improvement.

2 Qualities of Safety Leaders

Steve Farber used the acronym LEAP to discuss leadership in his address at the 2006 Professional Development Conference for the American Society for

Safety Engineers (ASSE).[8] He focused on the words: Love, Energy, Audacity, and Proof. By substituting "Data" for "Proof," I want to examine the qualities of leadership in terms of LEAD — a more memorable and meaningful acronym for remembering and teaching key leadership strategies.

Love and other "L" words

Leaders love their job, and love the people they lead, according to Farber. Of course, he's not referring to romantic love, but to relationships. Leaders respect and appreciate their followers, recognizing the integral role each plays in achieving the daily goals for the job they love. The "L" in LEAD represents other leadership qualities, as well:

Leaders attempt to *listen* actively, hearing both good and bad news. They put aside their biases and pay attention to everything in a communication. Before stating their viewpoint or opinion, they communicate respect for the speaker's words and emotions, and ask relevant questions.

Live, learn, love, and leave a legacy — Dr. Stephen Covey advocates these four hierarchical "L" words.[9] Empathic leaders learn the life phases of their followers; they know what consequences turn them on and which can be used to improve their work performance.

Workers at the "living" stage are "working to live," and want to receive fair financial compensation. All employees desire this, but some are also motivated by opportunities to learn. And through learning,

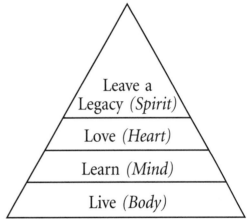

Four phases of living reflect motivating consequences.

these individuals get promoted to more challenging positions. Some learn to love their job and adopt the mindset of "living to work."

As people mature and consider the end of their lives, many contemplate their life accomplishments and wonder how they helped to make the world a better place. What could be more meaningful and emotionally fulfilling than working to prevent personal injury and saving lives? Safety leaders leave a legacy.

Energy and other "E" words

The subtitle of a book on teaching I recently edited with Phil Lehman is *Energy, Empathy, and Engagement in the Classroom.*[10] We derived these "E" words from a content analysis of the 39 essays in our book. The best university teachers — as with the best leaders — are *energetic* and *empathic*, and constantly search for ways to activate *engagement* among their students. The best leaders do the same with their employees.

Audacity and other "A" words

Audacity, according to Farber, means leaders "show a bold and blatant disregard for normal constraints in order to change the world for the better." He poked fun at the common slogan, "Think outside the box," by challenging the assumption there is a "box."

But safety standards constitute a "box," and performing outside the box implies at-risk behavior. Still, audacity is relevant for safety whenever leaders attempt to go beyond the tradition of engineering, education, and enforcement to increase energy and engagement in safety-related activities.

In safety there is a "box" of procedures and policies to follow in order to minimize the severity, exposure, and probability of injury. But there is also a "box" of safety procedures for maintaining compliance. This latter box is the one needing audacious, out-of-the-box thinking and acting. In this regard, two other "A" words are relevant: *avoidance* vs. *achievement*.

Almost every presentation at the 2006 Behavioral Safety Now conference referenced these "A" words.[11] Whether discussing leadership principles or intervention procedures, BSN speakers advocated a focus on *achieving* safe behaviors over *avoiding* at-risk behavior. In other words, audacious safety leaders think outside the enforcement box, and design interventions that put a positive, achievement spin on injury prevention.

"D" for data and demonstrate

Effective leaders align their behaviors with their values, setting an example for the action they want from their followers. By acting on our values, we demonstrate our commitment to both ourselves and others. Leaders who hold safety as a value consistently walk the safety talk. And if they don't, they experience a "guilt trip." (See "Power of a Guilt Trip" in Chapter Six, Understanding Personality.)

Data resonated at the BSN conference. First, every speaker had relevant credibility — proof of their expertise in behavioral safety.[12] And every intervention strategy was backed by relevant data.

I recommend using "D" for "data" instead of "P" for "proof." Then the acronym becomes LEAD.

Qualities of Leaders

- *Listen* actively.
- *Learn* continually.
- *Be Empathic* — Know the "turn-ons" and "turn-offs" of your coworkers.
- Leave a *Legacy.*
- *Energize* and *activate* Engagement in safety processes with your coworkers.
- Be *Audacious* — go beyond safety conventions, the traditions of engineering, education and enforcement.
- Put the focus on *Achieving* safe behaviors, not avoiding at-risk behaviors.
- *Demonstrate* your commitment — walk the safety talk.
- Use *Data* to demonstrate the effectiveness of your safety interventions.

3 Let's Get Real

In the previous article, I used the LEAD acronym to provoke group discussion about leadership:

L for Live, Listen, Learn, Love, and Leave a Legacy;

E for Empathy, Energy, Empowerment, and Engagement;

A for Audacity, and Achievement of success over Avoidance of failure;

D for Data to support an opinion or perspective.

Here I propose adding a critical A-word to this list — *Authenticity.* Specifically, leaders need to be authentic in their interpersonal relations.

What is authenticity?

My *American Heritage Dictionary* defines authenticity as "the condition or quality of being authentic, trustworthy, or genuine." This definition can incite constructive discussion about the meaning of related words: trust, reliability, consistency, and genuineness with regard to improving organizational safety. Even more behavioral direction is provided in two books with "authentic" in their titles — *Authentic Leadership*[13] by Bill George, former chairman and CEO of Medtronic, and *Authentic Involvement*[14] by the late Dan Petersen, a safety leadership guru if there ever was one.

The connection between these books is obvious: Authentic leadership yields authentic involvement. Let's review the primary authenticity directives provided in these books.

Authentic leaders

Authentic leaders "are more interested in empowering the people they lead to make a difference than they are in [having] power, money or prestige for themselves. They are as guided by qualities of the heart, by passion and compassion, as they are by qualities of the mind."[15]

Authentic leaders are vulnerable and always open to corrective feedback, and they demonstrate self-discipline to continuously improve. Bill George claims you cannot be authentic without compassion. Compassion is developed through profound understanding of other people's situations and feelings. Empathy is a synonym for compassion, and is a critical E-word for the LEAD acronym.

According to my dictionary, however, compassion also includes "the inclination to give aid or support or to show mercy."[16] People with empathy and compassion lead others with purpose, meaning, and personal values. They don't put an inordinate focus on short-run profits. They don't motivate through warnings and threats, which would de-motivate the development of self-accountability, a key component of authentic involvement.

Authentic involvement

Authentic involvement is self-directed, and occurs when people are "treated like a mature, adult human being; as an equal, not subordinate, able to use their innate intelligence and skills daily, even hourly; able to achieve; given responsibility; and recognized for doing a good job."[17]

So who treats employees like this? You know the answer — authentic leaders. Actually, all of the leadership principles reflected in the LEAD acronym are relevant here. Effective leaders enrich their work culture and help workers become self-directed, self-accountable, and self-motivated. Dr. Petersen advocates an integration of the humanistic and behavioristic approaches to understanding and helping people. This is, in fact, the foundation of People-Based Safety™, which I call "humanistic behaviorism."

Petersen advocates shared decision-making between salary and hourly workers, with each side recognizing the need for interdependent cooperation. But for this to happen, managers, supervisors, and hourly workers need mutual training on effective problem-solving tools and methods. He suggests training on specific analysis techniques, such as statistical process control (SPC) which includes the use of fishbone diagrams, pareto charts, flowcharts, control charts, and scatter diagrams.

Problem-solving mechanisms

Dr. Petersen also discussed various ways to enable regular employee input on safety-related matters to facilitate authentic involvement:

Safety Improvement Teams — Management asks employees to address a specific safety issue.

Job Safety Analysis — Work groups define specific environmental and/or behavioral hazards associated with each step of a job and develop ways to eliminate or control them.

Hazard Hunt — Employees use a special form on which to report anything they feel is a hazard, followed by corrective-action feedback from management.

Ergonomic Analysis — After training on ergonomic principles, workers observe the various behaviors of a job and consider ways to decrease the probability of a cumulative trauma disorder.

Incident Recall Technique — Through one-on-one interviews, employees relate a specific close call they experienced or heard about, suggest contributing factors to the incident, and then explore ways to prevent similar incidents and potential injuries.

Observation and feedback

I would add the observation and feedback process of behavior-based safety to this list, including employees' development, application and refinement of a critical behavior checklist (CBC). Workers use this CBC to coach each other, which includes observing safe vs. at-risk behaviors on the job, defining barriers to safe behavior and facilitators of at-risk behavior, and providing constructive behavioral feedback to the worker.

None of these employee-involvement techniques can work without sufficient management support and authentic leadership from both supervisors and workers. This is easier said than done. Example: What happened to those safety-suggestion boxes once found at most industrial sites? Answer: They were not used

3.2 – A CBC facilitates one-on-one coaching.

constructively and, consequently, were removed because of insufficient leadership among managers and hourly workers.

How can I help?

In a culture rich in safety values, everyone at every level of the organization strives to be authentic in his or her own way. Instead of asking "Who, what, or why?" after an incident, they ask "How can I help?"

How can we be more authentic? Facilitate group discussions around behavioral definitions of "authenticity," with particular reference to industrial safety. Then draw up a list of what it will take to "walk authentic talk" and improve safety.

"Real" Leadership Qualities

- Be *open* to feedback.
- Show *compassion*.
- Exert *self-discipline*.
- Show your willingness to be *vulnerable*.
- Demonstrate a *sense of purpose*.
- See the *big picture* — don't focus exclusively on short-term issues.
- Help coworkers become *self-directed, self-accountable* and *self-motivated*.
- Give *responsibility* and *recognize* a job well done.
- Treat adult coworkers as *mature* adults.
- Ask, *"How can I help?"* improve safety.

4 How Are We Doing?

The "D" of the LEAD acronym stands for "Data." Data both directs and motivates behavior. By observing the results of our actions, we learn how well we completed a task and what we can do to improve.

But some data are useless, misleading, and de-motivating. For example, injury statistics based on self-reporting are unreliable and have no diagnostic value. And they can activate distress or a false sense of security. Leaders need to use data strategically to direct and motivate themselves and others.

Accountability data

"What gets measured gets done." This popular slogan reflects the connection between data and accountability. But using wrong data to assess accountability can be disastrous. "What could be worse," asked Dr. Edwards Deming, than "holding willing workers accountable for numbers they cannot control?"[18]

Dr. Deming taught us the critical difference between behavior and performance, a distinction needed to select and examine the right data. Many behavioral researchers and safety professionals use these words interchangeably, but my online dictionary (www.m-w.com) defines performance as "something accomplished" and behavior as "the manner of conducting oneself."

This behavior/performance distinction is critical for giving the right kind of feedback. Specifically, when can we hold people accountable for data? The answer is simple: Hold people accountable for data they directly influence.

In safety, it's fair to hold ourselves accountable for the variety of activities we can do to prevent personal injuries — from coaching others regarding their safe versus at-risk behaviors to completing hazard recognition and close-call reports. Likewise, if an individual's behavior or lack thereof is clearly linked to an injury, it is legitimate to hold that person accountable (in part) for the performance data reflected by injury statistics. But the contribution of environmental factors beyond the individual's control should be acknowledged.

Some performance deficits result from *behavior* deviating from the process. But performance deficits also occur from *system factors* (physical conditions, management decisions) independent of process-related behavior. Hold people accountable for the first, but not the latter.

Isn't this common sense? Then why does there seem to be so much emphasis on injury statistics or performance data at safety meetings? How often is a graph of safety-related behavior displayed to illustrate accomplishment (or failure) at injury prevention? Bottom line: Show

Process data reflect desirable vs. undesirable behaviors.

process data to individuals and groups that reflect their controllable actions associated directly with performance data.

Leadership data

Almost every book on leadership presents information on the person characteristics of leaders. For example, the recent text by Dr. Thomas Krause, *Leading with Safety*[19] connects leadership with five personality traits — emotional resilience, extravorsion, learning orientation, collegiality, and conscientiousness. Dr. Krause also distinguishes between transactional leaders (or managers) and transformational leaders with certain interpersonal styles (including challenging, engaging, inspiring, and influential). In this book, I describe leaders as individuals who are energetic, passionate, open, trust-worthy, compassionate, goal-directed, self-confident, intelligent, and flexible.

It's fascinating and entertaining to explore one's personality, and consider the correlation between specific person factors and behavior. Many readers have taken the Myers-Briggs personality inventory,[20] and enjoyed learning about the behavioral implications of certain person qualities and styles. (For more on the correlation between personality and leadership, see Chapter Six, Understanding Personality.)

I urge caution when considering these data. First, the assessment tools for personality data are often unreliable and invalid. Secondly, the connection between most person data and behavior is ambiguous or weak. But the critical issue is applicability.

Can data suggesting leadership-related personality traits, states, or styles provide directional or motivational feedback to an individual? Actually, using these data to influence ourselves or others is analogous to developing an action plan from an organization's injury data. In both cases, the data are unreliable and influenced by undefined factors independent of people's behavior. And neither provides useful diagnostic information to direct continuous improvement.

Dr. Krause acknowledges low practical value in assessing leadership-related characteristics of people. Telling people they score high or low on a measure of charisma gives minimal direction for improving leadership.

But to the extent it's possible to define a particular leadership quality in terms of specific behaviors, personality data can be useful. For example, by observing people judged to be charismatic, it might be possible to identify behaviors that reflect this label and then tell people what they can do to demonstrate charisma. Subsequently, a person can be observed and given behavior-based feedback related to the presence or absence of charisma-related behaviors.

Drs. Aubrey and James Daniels advance an entirely different perspective in their book, *Measure of a Leader*.[21] They claim the measure of a leader should focus on the behavior of the followers. The key type of follower behavior: "discretionary behavior" supporting the leader's vision.

This is behavior that exceeds a worker's job requirements. It is self-directed, meaningful, and intrinsically reinforcing. I refer to this type of behavior as "actively caring" whenever it relates to injury prevention or health promotion.

Increasing discretionary behavior

The Daniels brothers focus on the appropriate use of "positive reinforcement" to increase discretionary behavior. With threats and punitive

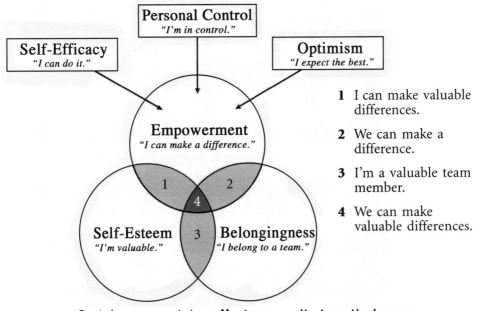

Certain person states affect propensity to actively care.

consequences, people do not become self-accountable, and do only what's required.

The approach I've advocated for increasing actively-caring behavior is consistent with these suggestions. Research indicates people are more likely to help others (or emit discretionary behavior) when they have relatively high levels of self-esteem, self-efficacy, personal control, optimism, and a sense of belongingness.[22]

Genuine behavior-based rewards and recognition are likely to enhance these person states. But there are other ways to facilitate the occurrence of these person states and increase the probability of discretionary behavior. (See Chapter Four, Enriching Your Culture.)

One final point: Please be skeptical of people's opinions, even if they sound like good common sense. I recommend frequent use of the phrase, "Where's the data?" And when someone shows you data, ask another question, "How can these data be used to facilitate continuous improvement?"

Leadership Principles

L for Live, Listen, Learn, Love, and Leave a Legacy;

E for Empathy, Energy, Empowerment, and Engagement;

A for Audacity, and Achievement of success over Avoidance of failure;

D for Data to support an opinion or perspective.

How Leaders Develop Accountability

- Performance is "something accomplished."
- Behavior is "the manner of conducting oneself."
- This behavior/performance distinction is critical for giving the right kind of feedback.
- Hold people accountable for data they directly influence.
- Some performance deficits result from *behavior* deviating from the process.
- But performance deficits also occur from *system factors* (physical conditions, management decisions) independent of process-related behavior.
- Hold people accountable for the first, but not the latter.
- Show process data to individuals and groups that reflect their *controllable actions* associated directly with performance data.

5 The Challenge to Avoid Accepting Failure

The winds of grace are always blowing, but you have to raise your sail.
— Ramakrishna

Safety professionals strive for "injury-free." Their ultimate outcome statistic is "zero." Similarly, after treatment for prostate cancer, the desirable prostate-specific antigen (PSA) score is "zero." A post-surgery PSA above 0.001 indicates the presence of cancer. The patient is not cancer-free.

After surgery for prostate cancer in 2002, and for five consecutive PSA tests, my score was 0.001. I was cancer-free. After three of these ideal test results, my urologist's office only scheduled me annually for a PSA test. Time passed and I realized it had been a year since my last assessment, and I wanted verification of "cancer-free."

I could not have been more optimistic as I waited for my test results. The doctor was friendly and cheerful as he grabbed the computer printout from the holder on the door. Suddenly, his demeanor changed dramatically. I knew something was wrong. He put his chewing gum in a paper towel, and with a look of surprise, he delicately told me my PSA was 0.18.

While 0.18 is close to zero, I knew this was not good news. It means cancer is in my body, and can certainly grow to be debilitating and life-threatening. Prostate cancer is the second-leading cause of cancer-related deaths in the U.S. The urologist reviewed my records and noted how unusual it was to see cancer after surgery and post-operative assessments as successful as mine. Such rhetoric did not make me feel any better.

Then he surprised me with the following: "At this time, I see no need for radiation, chemo, or Lupron." (Lupron is a last-resort drug that acts like a female hormone and can have a number of undesirable side effects.)

The doc asked me to schedule a reassessment of my PSA in four months. "In the meantime, you might check the Internet for possible advantages in taking vitamin E, lycopene, selenium, and vitamin D." This is neither new nor useful information, I thought to myself. I've taken these supplements for years, even *before* my cancer diagnosis.

Safety glasses? Check. Sterile gloves? Check. Now please be Actively Caring down there.

An unsuccessful surgery can fuel failure acceptance.

Relevance to safety

Do you see the relevance here to safety leadership? Obviously, safety leaders like to "keep score" and strive for a score of zero incidents. Whether targeting cancer-free or injury-free, anything higher than zero is discouraging and could have dire consequences.

With this scoring system and a mission to remain at zero, it is easy to adopt a failure-avoidance mindset. Just avoiding failure becomes the focus of interpersonal conversation and self-talk. This can detract from necessary attention to upstream actions needed to maintain zero. Plus, we feel less distressed and more empowered when working to achieve as "success seekers" than when working to avoid failure as "failure avoiders."[23]

Fighting complacency

I got pretty comfortable with my post-operative PSA scores, and so did my surgeon and his staff. When I left the doctor's office to schedule another appointment, the office administrator greeted me with, "I'm so glad you're doing so well." She hadn't seen my discouraging PSA, and expected only the best.

Does this happen in safety? Can a score of zero (for injury-free) lull leaders into a false sense of security? Without a success-seeking mindset and mindful attention to daily activities needed to remain injury-free, leaders can get complacent about safety. "We've been injury-free for two years, so if we keep doing what we're doing, we will maintain this ideal score." No need to change — no need to learn more about injury prevention — no need to worry.

Fighting failure acceptance

Safety leaders don't view a workplace injury with this mindset, do they? That's tantamount to accepting occurrences of injuries until they reach some critical number. Just keep doing what we're doing until our injury rate exceeds the standard for our business — of course, that's not acceptable.

But have you ever been at a loss for what to do in order to return to zero, to injury-free status?

Success seekers need strategies

My doc offered no new strategy for returning me to cancer-free. From my perspective, my behavior had not changed.

Was the return of my cancer only a matter of bad luck? Was it merely genetically determined (a common view of the urologists I've interviewed)? Must I accept the perspective, "Not all cancer can be prevented?" If not, then what should I do?

Are all injuries preventable? Are some injuries the result of uncontrollable factors? Should we keep on doing what we're doing, and accept failure?

Of course not; but sometimes safety leaders need to ask my cancer-related question, "What should we do?"

Raise your sail

In 2005, I gave the closing general-session keynote at the 21st Annual Voluntary Protection Programs Participants' Association (VPPPA) Conference in Dallas, Texas. A crowded room of about 2,000 listened intently as I introduced the audience to "People-Based Safety™" — an evolution of behavior-based safety. They reacted to my concepts, stories, and cartoons with passion and energy.

"We need to make safety personal," I said. "Face the brutal facts of injuries and their contributing factors, build trusting relationships with people, and actively care with an interdependent mindset." Then, without any forethought, I told my cancer story — emphasizing the disheartening news I had received just five days earlier.

Receiving inspiration

This turned out to be "my most inspiring" keynote — not because of anything I said, but thanks to how the audience inspired me. I was touched by the many caring emotions displayed on the faces of the attentive listeners. After a standing ovation, I was swamped by an onslaught of participants expressing encouragement, wishing me good health, and telling me I'll be in their prayers. Many asked for my email address so they could send information relevant for my fight with cancer, including the names of cancer specialists I could contact. Two participants waited more than 30 minutes to privately pray for me.

Special people

I left the ballroom 45 minutes after my keynote feeling inspired and more ready than ever to fight for zero again. That special group of safety leaders lifted my self-esteem and optimism higher than I thought possible in my current state. In a world that seems filled with mistrust, conflict, selfish entitlement, and debilitating fear, it's uplifting to see so much genuine caring from so many people.

What's the safety lesson here?

First, I'm convinced safety leaders are a special breed of people gifted with emotional intelligence. They are willing and able to share with others. If only those "others" would ask. Plus, safety leaders have knowledge, practical tools, and procedures they can help others use for injury prevention. If only those "others" would ask.

Safety conferences like the VPPPA are so rewarding for safety leaders. They get to interact with like-minded people who want to share and receive strategies to achieve and sustain an injury-free workplace. They return home inspired, possessing new ways to keep people safe. I hope their workplaces are receptive. But if they are not, I'm confident safety's leaders will not accept failure and become complacent. They will inspire those "others" to fight the good fight. That's what they did for me.

Your role

People-Based Safety™ leaders should help ensure the public never forgets the calamitous destruction and human misery that occurred because appropriate preventive action had not been taken in New Orleans. With billions of dollars, New Orleans may be rebuilt and protected with the proper levee system, but the money can't bring back the hundreds of lives lost nor restore the historic structures, memorable possessions, and priceless keepsakes washed away by flood waters.

I hope we never let people forget that most of these losses could have been prevented with the types of proactive plans and activities advocated daily by safety leaders. And may we find consolation and encouragement in observing that many people will react to tragedy by actively caring, as many did for those victimized by the hurricanes of 2005.

CHAPTER FOUR

Enriching Your Culture

Introduction

You might have noticed the subtitle to *Leading People-Based Safety™* reads "Enriching Your Culture."

You simply can't do it without leading people. In Part 1 of this chapter, Dr. Geller specifies 10 strategies for powering participation — activating your workforce's efforts in injury prevention. All of PBS's ACTS components come into play:

• You *Act* to motivate with positive consequences;

• You *Coach* using SMART goals (Specific, Motivational, Achievable, Relevant, and Trackable) and empowering, positive language;

• You get people to *Think* about safety in personal terms, their relationships with coworkers, not hard comp costs and cold group statistics; and,

• You teach and motivate with personal storytelling by victims and incident witnesses in order to activate vivid images and allow workers to *See* what is truly at stake.

Parts 2 and 3 cover the important leadership actions of giving rewards and recognition when warranted, and corrective actions when called for. Be careful here, for leaders can go astray and cause more harm than good by miscalculating how to reward safe performance, and how to correct at-risk performance. As Dr. Geller says, "Avoid the cultural impetus to use nonproductive and quick-fix punishment, and eventually you might see beneficial change in your culture."

Cultures involve many group activities, don't they? Part 4 shows the damage that can be caused when group behavior goes bad. Because groups often operate with diffused responsibility, a lack of personal identity, and what psychologists call "pluralistic ignorance" or "groupthink," their actions can have a negative emotional impact on people's morale, self-esteem, sense of ownership and belongingness, and their willingness to actively care — all attributes of a rich culture.

What to do to counter groupthink? Dr. Geller's ideas include "Don't silence disagreement" and "Don't rush to quick group decisions."

Parts 5 and 6 are case studies in real-world cultures — not organizational cultures, but studies of social cultures that demonstrate the need for important leadership qualities, such as empathy, openness, perceptiveness, and appreciation of diversity, in order to sustain culture enrichment.

Part 7 closes this chapter on enriching your culture with a heart-rending true story of a group of friends and family members who come together to courageously support and care for a man facing an enormous health challenge, an "accident" of fate over which he had no control. Together the acts, beliefs and values of this group comprise an informal culture. They show that we can't enrich lives, or a culture, without a lot of caring, courage, sensitivity, and interdependence. We certainly can't do it alone.

Dave Johnson, Editor
Industrial Safety & Hygiene News

1 Ten Strategies to Power Participation

I'd like to offer 10 prime ways to get people involved in injury-prevention efforts — and enrich your Total Safety Culture in the process. Safety leaders might find it useful to have this Top 10 list in one place. Perhaps because you are a leader, these strategies will strengthen your resolve to achieve more employee participation in your injury-prevention campaigns.

10) Make safety personal

How do successful advertisers sell their products? They display individuals similar to their potential customers enjoying the benefits of their products. In contrast, many organizations try to motivate safety involvement with group statistics like total recordable injury rate (TRIR) or workers' compensation costs. This takes the focus away from what people relate to — other people.

9) Teach and motivate with personal stories

A well-told, personal story activates vivid imagery. Listeners can put themselves in the position of the storyteller and feel relevant emotions. When the story is linked to a related lesson, learning is facilitated and remembered.

8) Accompany scare tactics with action plans

A personal story about an injury, pain and suffering is emotional and motivational. Listeners visualize themselves in the same predicament, and experience fear and anxiety. They are ready to take action to prevent such a tragedy in their own lives. This is the prime time to teach an injury-prevention technique.

7) Activate and support success-seeking

Scare tactics and prevention strategies activate desirable behavior, but can also lead to an undesirable attitude or mindset. When the focus is on avoiding failure, one's sense of personal control and freedom is stifled. And if the prevention efforts do not work, you can get failure acceptance, apathy, and feelings of helplessness.

The antidote: Substitute success-seeking for failure-avoiding. We should define our injury-prevention efforts and results in achievement terms. Get people talking about what they *do* for safety, and discuss outcomes in terms of milestones accomplished instead of losses avoided.

6) Motivate with positive consequences

In so many situations, attention to failure rather than success is more prevalent, and penalties are used more than rewards to manage the behaviors of children and adults. Why? Because mistakes are noticeable, and punishment seems to work.

But recognition for desirable performance is key to a success-seeking attitude. So, as trite as it sounds, become a "good finder." Look for the good things people do and support that behavior with positive consequences. But remember the power is in the delivery. Sincere one-on-one words of genuine appreciation are usually more influential than financial rewards.

"Good finders" help others
deal with failure.

5) Focus on the process

In safety, the focus is typically on negative outcomes, from OSHA recordables to compensation costs. With our attention on the negative and reactive scoreboard of total recordables, it's easy to take our eyes off the "ball" — the proactive process things we need to do daily in order to prevent workplace injuries.

For more than a decade, my Safety Performance Solutions partners and I have been teaching corporate cultures a DO IT process, with "D" for define desirable or undesirable behavior to target; "O" for observe the target behavior; "I" for intervene to improve the behavior; and "T" for test the impact of your intervention. This process defines the journey needed to come closer to the ultimate destination of an injury-free workplace.

4) Use behavior-based feedback

Behavior-based feedback provides direction or motivation, or both of these, depending on its delivery. I only want to make the point that behavior-based feedback is necessary for performance improvement and competence building. Since people want to be more competent at what they believe is important, opportunities to receive feedback invite participation.

3) Set SMART goals

"Zero injuries" reflects the vision of dedicated safety leaders. It is a destination, not a goal. Goals are journey tools.

SMART goals are empowering because they facilitate a process that people believe is achievable, effective, and worth the effort. This is reflected in the words represented by the letters of SMART: Specific, Motivational,

Achievable, Relevant, and Trackable.

2) Use empowering language

For more than three decades I have been complaining about unfortunate language used by safety leaders. Words like "behavior modification," "accident investigation," "loss control," "compliance training," "root cause," and "occupant restraint" come across as failure-oriented and freedom-limiting. They set the stage for finding faults rather than getting at the facts.

Our language both influences and reflects our culture. Take a careful look at your safety language and make necessary

S pecific

M otivational

A chievable

R elevant

T rackable

SMART goals are empowering.

adjustments. This is especially important for safety leaders of all stripes. You set the tone for your team, department, shift, organization, etc.

For example, substitute "incident analysis" for "accident investigation," "contributing factors" for "root causes," and "safety belt" for "occupant restraint."

Perhaps if we stopped calling it a crib he would be happier...

EXECUTIVE SUITE

Language affects attitude and vice versa.

1) Ask the right question

Consider the effects of questions like "Who did that?" "Why was that hazard not removed earlier?" "What is the root cause of this injury?" and "Why didn't you follow the OSHA guidelines?"

These questions project the problem beyond the person asking the question. They deflect a solution to someone else.

The perspective changes with this question, "How can I help?" It's not "Why did that have to happen?" but rather "What can I do to help correct the mishap?" I recommend leaders start with the question, "What can I do?" and proceed to do whatever is within your domain of influence to focus on the positive and make your injury-prevention process more achievement-focused.

2 Reinforcement, Reward and Recognition

Enriching your Total Safety Culture will necessarily involve acts of reinforcement, reward and recognition. Here I want to make sure you understand the critical distinctions between positive reinforcement, reward, and recognition, and explain how the standard behavior-based safety (BBS) instruction for giving recognition can be undesirable.

Listen to a BBS trainer or consultant and you'll likely hear the term "positive reinforcement." Many BBS trainers claim positive reinforcement is the most effective procedure for improving safety-related behavior.

Misuse of technical language

Trainers and users of BBS throw around the term "positive reinforcement" too freely. My partners at Safety Performance Solutions (SPS) and I use the term "reward" instead. Here's why: Reinforcement is a procedure using consequences to increase behavior. If the target behavior does not increase in *frequency, intensity,* or *duration,* the procedure was not reinforcement and the consequence was not a reinforcer.

Plus, only behaviors are reinforced, not people. It is simply incorrect to say, "I want to reinforce you for actively caring."

In the workplace, it's usually impossible to observe whether a consequence is a reinforcer. In fact, positive consequences are typically delivered when a person goes beyond the call of duty — meaning the target behavior is already at a high level of frequency, intensity, or duration. Therefore, these consequences only serve to support or maintain behavior, not reinforce it.

It's better to reward

Avoid confusing the technical versus dictionary meaning of "reinforce" and "reinforcement."

How?

Simply use the term "reward." Rewards are directed at people, with the intention of improving or maintaining their desirable behavior. Some rewards are given long after the occurrence of the target behavior. Some are not even associated with behavior, but rather reflect a series of achievements from an individual, team, or entire organization. In many situations rewards do not directly influence behavior.

That's OK. Rewards can make a person feel better, and this is a worthwhile outcome by itself. Plus, when rewards increase such internal person states as self-esteem, personal control, or optimism, they have beneficial indirect impact on desirable behaviors.

These person states increase a person's willingness to look out for the safety of others — a critical element to enriching your culture. Whether or not a reward increases the behavior it follows, it is apt to improve one or more feeling states that make people more likely to actively care for other people's welfare.

It's all in the delivery

The presentation of a reward can be more influential than the material consequence. Rewards are not payoffs for performance, but rather a means of recognizing people for their special efforts. How can you associate emotional significance with a reward? You got it — it's in the delivery.

More than 20 years ago, the best-selling *One-Minute Manager* by Ken Blanchard and Spencer Johnson[1] urged us to deliver one-minute praise every day to the people we manage, along with one-minute goal-setting and one-minute reprimands. Today, we say goals need to be SMART (for Specific, Motivational, Achievable, Relevant, and Trackable). We also claim praise or interpersonal recognition should occur far more often than reprimands, by at least a 4-to-1 margin. Plus, we substitute "corrective action" for "reprimand," and incorporate empathic listening and interpersonal coaching.

The strategy for giving interpersonal recognition has not changed much over the years, except some claim recognition should be public, while others claim recognition should be delivered privately and one-on-one. I recommended private over public recognition. Why? Because some people feel embarrassed when group attention is directed toward them. Sometimes those recognized fear negative consequences from peers, perhaps because of envy or because it's not "cool" to be praised for safety.

Public recognition can also be de-motivating to members of the audience who believe they also deserve the reward. Public recognition of five individuals at the staff Christmas party of a large construction firm illustrates my point.

For the first time in its 25-year history, the owner and CEO handed recognition plaques to five employees for "going beyond the call of duty." These five individuals were relatively new hires, and many in the audience of 300 looked at each other in dismay: "Why should those five short-term employees get this rare recognition over many others who have done so much for our company over a much longer term?"

"Mischievous sarcasm"

For more than two decades, I have taught safety leaders specific sequential steps for giving recognition. In his 2005 book *Praise for Profit*,[2] Jerry Pounds describes how he and colleagues at a large behavior-focused consulting firm taught the same basic recognition steps for more than 30 years. We both taught participants to specify the target behavior, deliver it soon after you observe the behavior, be genuine and personal, use "I" statements, resist bringing up other matters, and relate the behavior to a higher-order quality like leadership, integrity, or trustworthiness.

This sounds good, right? Well, Jerry Pounds offers a reality check. He says thousands of supervisors who attended his classes on behavior-based recognition "rebelled vehemently" to following the recognition steps. Many complained "they would feel stupid and out of character going around saying nice things to employees after years of ignoring them."[3]

As a result of this apparent "manipulation," Pounds reports, "what we were asking the supervisors to do was humiliating for them and the employees." As a result they "adopted an attitude of mischievous sarcasm" with comments like "I hate to do this but I'm going to have to reinforce you for that." Later, many employees used the recognition process as an excuse to escape accountability. "If their performance did not reach expectations, they said it was because they had not been appropriately praised for their efforts."[4]

Keep it simple

Dale Carnegie had it right. The best and simplest way to recognize people is to show genuine interest in what they are doing.[5] Similarly, Dr. Blanchard has recently simplified the responsibility of the "one-minute manager" to convincing people they are doing worthwhile work.[6] So we return to my prior point about meaningfulness and delivery.

Effective recognition is authentic.

A simple "thank you" can be a powerful support of desirable behavior and a booster of self-esteem and optimism. But this statement of gratitude must be sincere and genuine. Plus, the recipient knows what action(s) warrant the praise and believes this action reflects worthwhile work.

Bottom line: Set aside that sequence of behavioral steps you were taught for giving interpersonal recognition, and merely show bona fide interest and appreciation in what people do to keep themselves and others safe. You might not increase the frequency of behavior already high in the person's work priorities. But you will likely enhance positive feelings about the job, leading to a valuable boost in self-worth, competence, and a sense of belongingness. And those feelings, when shared by individuals throughout your organization, will enrich your Total Safety Culture.

Leading with Rewards

- Use the term "reward" in place of "positive reinforcement." Only behaviors are reinforced, which can sound cold. Rewards are directed at people.
- Be careful: Public recognition can de-motivate members of an audience who believe they also deserve the reward.
- Be sure your reward or recognition presentation does not come off as manipulation.
- Don't over-kill rewards and recognition. A simple "thank you" can be a powerful support of desirable behavior and a booster of self-esteem and optimism — if it is genuine.
- Keep it simple: Simply show bona fide interest and appreciation in what people do to keep themselves and others safe. You might not increase the frequency of behavior already high in the person's work priorities. But you will likely enhance positive feelings about the job.

3 Punishment to Fit the Crime?

As some of you might know, I'm a huge Hokies (Virginia Tech) football fan. Some would say fanatic. A few years ago, at the Gator Bowl, the quarterback for Tech stomped his foot on the leg of the player who had tackled him.

Following a chorus of criticism, including the Virginia governor-elect saying the incident "made my heart sink," the quarterback (who led Virginia Tech to an 11-2 season) was dismissed. Initially, the punishment, as determined through conversations with the University president, athletic director and football coach, was to be a two-game suspension next season.

But suspension was changed to dismissal when these University officials learned the quarterback had received a speeding ticket weeks before the bowl game for driving 35 miles per hour in a 25 mph zone sometime after 2 a.m. The star quarterback already had been suspended from the team for the entire 2004 season because of a conviction of reckless driving and possession of marijuana. In the context of a last-chance opportunity, the speeding ticket was apparently the last straw.

"It's the culture, stupid!"

I borrowed the above subheading from several talks I've heard at safety conferences, and it's so relevant here. Our culture looks for quick-fix solutions to its problems. And no solution is as swift and efficient as punishment.

Corrective action should acknowledge desirable behavior.

Will we ever own up to the ineffectiveness of punishment as a corrective measure?

Will we ever put more attention on proactive prevention than reactive retaliation?

Will the unsung heroes who provide critical proactive intervention, from school teachers and social workers to safety leaders, ever get the credit and financial support they deserve?

Will we ever actively care beyond our own soon, certain, and significant consequences?

It's the way our culture operates. And if you sincerely want to enrich your organization's culture, be very careful with your use of punishment, penalties, enforcement and discipline.

Where's the corrective action?

The quarterback was advised to express sincere regrets to the player he stomped. But newspaper accounts of the incident report the other player did not make himself available for an apology, and was quoted as calling the quarterback a "no-class individual." This is all secondhand information, and many details are missing, but this is what the public saw. Question: What corrective action had the quarterback received since his serious troubles in 2003?

Here's another question: Was professional counseling provided for the quarterback? Indeed, is any corrective intervention other than punitive consequences for misbehavior given athletes who need to improve their emotional intelligence?

At a 40-minute press conference, our University president, athletic director, and football coach discussed the decision to impose the ultimate punishment on an athlete who needed another year of NCAA football to achieve a top ranking in the NFL draft. No commentary was offered that even remotely related to corrective action — past, present, or future. And none of the news reporters posed a question relevant to this critical issue.

While the media reflects culture, it also influences culture. This media event was a disheartening exemplar of our quick-fix reactive culture, giving no attention to corrective action.

Positive enrichment

There is a way to enrich your culture through adversity. I know the safety director of a construction company where a worker was killed on the job. She is determined to gain safety-related improvements from this accidental death. How? She is attempting to make every conversation about that awful event, even those with OSHA inspectors, include something constructive that could be done to prevent another fatality or serious injury at the variety of construction sites managed by her company. The fatality increased management and workers' commitment to safety. Now she is focused on increasing the competence of the company at remaining injury-free — one day at a time.

Bottom line: Destructive events like those reviewed here increase motivation to improve from everyone involved. This is a prime time to learn how to prevent related mishaps and implement preventive intervention. Avoid the cultural impetus to use nonproductive and quick-fix punishment, and eventually you might see beneficial change in your culture. As a safety leader, you know your example can make a difference.

Leading with Corrective Action

When it comes to destructive incidents — injuries or fatalities — don't opt for quick-fix fault-finding and punishment. Observe how your organization might mirror our society's quick reactions to "assigning blame" and "fixing problems." Avoid this cultural impetus. Use misfortune as an opportunity to increase safety competence. It starts with the example you set.

4 When Group Behavior Goes Bad

I want to discuss a dynamic that is positively detrimental to our objective of enriching your organization's culture. As a safety leader, you need to be aware of its destructive force.

My 12-mile bike ride one morning was interrupted by an event that adversely affected my attitude and self-talk. Here's what happened: At the beginning of my route near my home in Newport, Va., I encountered about 20 bikers traveling in the opposite direction. All were decked out in radiant racing attire, consistent with their sleek road bikes. Of course, I was using the appropriate safety gear — personal protective equipment, if you will — including bike helmet, gloves, and safety glasses, as seemed to be the case for all bikers in the group.

**Groups can facilitate
bad behavior.**

As I approached the group, one of the bikers yelled out with a condescending tone, "Look what we've got here." I heard a few chuckles, and then another group member barked, "You need a new bike." My reaction: "You mean like the one your daddy bought you? And do I also need one of those girly shirts?"

Two factors stopped me from verbalizing these negative comments aloud. First, I was startled by their negative verbal attack, and before I could think of a clever retort, the bikers were gone. And these derogatory comments came from about 20 individuals, each younger and more fit than I.

Emotional impact

It should have been so easy to blow off those comments. After all, I understand the psychological principles behind the deprecating comments. The cyclists likely expected social approval for their outbursts. And responsibility for the criticism could be diffused among other members of the group. That diffusion of self-accountability was facilitated by the group's homogeneity.

But explaining behavior does not necessarily minimize its impact. I ruined the first half of my bike ride with negative and destructive self-talk such as:

"Why is the younger generation so confrontational?"

"Why didn't a lone biker twice their age or more get waves of social support instead of ridicule?"

Missed opportunity

Engrossed in my negative self-talk and pedaling up a long and steep hill, the group I saw earlier was now dispersed and seemingly racing individually down the hill to a finish line. Every biker waved enthusiastically, and many shouted positive words in a supportive tone.

"Good morning."

"It gets easier."

I just nodded my head and pedaled on. That brief encounter 40 minutes earlier suppressed positive emotions and caused me to miss an opportunity to reciprocate encouraging hurrahs.

Group vs. individual decisions

Those positive statements from individual bikers contrasted sharply with the earlier negative remarks from the group. Let's consider three factors that contribute to this behavior in a group versus a one-to-one situation.

1) *Diffused responsibility*: It is natural for people to feel they cannot be held personally accountable when taking a risk — such as shouting critical remarks at a passerby — as part of a group or team. The risk of failure (or any sort of retaliation) is spread around.

How often does a work group decide to bypass or overlook a safety protocol, perhaps for more efficiency or productivity? If this results in someone getting hurt, no one person can be held accountable. The risk was a group decision.

2) *Deindividuation*: Deindividuation — loss of identity and sense of personal responsibility — is facilitated when group members wear uniforms and cut their hair in similar ways, as in prisons, cults, monasteries, and the military. The bikers I encountered all were dressed in similar apparel and rode high-end racing bikes.

Generally, whenever a group fails to support individual contributions, group members might give up their identity and sense of personal responsibility. This can lead to a perceived loss of personal control and contribute to complacency.

3) *Groupthink*: Deindividuation can influence some people to avoid speaking up when they disagree. Silence is interpreted as consent and supports an illusion of group unanimity. Social psychologists call this "pluralistic ignorance."

I bet you can think of several examples of groupthink. Whenever groups attempt to reach a quick decision without substantial

Homogeneity stifles diversity and synergy.

discussion, the probability of groupthink is increased. Whenever the leader of a group stifles disagreement and advocates unanimity, it risks the disadvantages of groupthink.

Of course, a group leader who embraces diverse opinions, invites input and critique, and challenges individuals to think outside the box decreases the probability of groupthink. Also, effective leaders do not let negative emotions impact their actions. Obviously, I was not emotionally intelligent that morning. I hope reviewing the principles discussed here will help me do better next time. How about you?

Deflating Groupthink

- Don't stifle disagreements within your group or team, or at a safety meeting.
- Don't rush your group for a quick consensus or decision.
- Embrace diverse opinions.
- Invite input and critique.
- Challenge individuals to think "outside the box."
- Don't let your emotions get the best of you when there is adversity or resistance coming from your group.

⑤ Case Study: The "Nickel and Dimed" Culture

This is the first of three case studies examining different types of cultures. All of my examples come from outside the domain of safety, but through the study of other cultures I believe leaders learn how to enrich their own culture.

"Nickel and Dimed"

I must walk through Atlanta's Hartsfield International Airport 20 times a year. But I never noticed that booth vendors have no place to sit. That was before I read *Nickel and Dimed* by Barbara Ehrenreich.[7] Now I have a new appreciation for working conditions of all sorts.

Take the woman selling sunglasses in the Atlanta airport. Sitting in a waiting area one time, I watched her for 30 minutes. She made no sales but merely stood calmly in front of her booth.

I approached her and introduced myself.

"How long did you work today?"

"Eight hours."

"Did you get tired standing?"

"Yes, very much."

"Why don't you have a chair or stool to sit on?"

"We're not allowed."

"How much do you make per hour?"

"$7.50."

"How old are you?"

"38."

Then she volunteered she has only been in the U.S. for two years, and asked if I could do something about the "no-sit-down rule."

"If only I could sit down once in a while, I wouldn't be so tired."

Simple changes in working conditions can mean a lot.

The low-wage world

I must confess I've never conversed with an airport vendor like that before. But *Nickel and Dimed* has changed my view of both the work world and my own behavior. From 1998 to 2000, Ehrenreich took a number of hourly jobs in three states to study the culture of the low-wage work world and to see how well she could live as a minimal-wage worker. She worked as a hotel maid, house cleaner, and nursing-home aide in Maine, a Wal-Mart sales associate in Minnesota, and a waitress in Florida.

She had to work two jobs at $6 to $7 per hour to cover the cheapest lodging available. And none of her jobs provided overtime pay, retirement funds, or health insurance. Plus, she found that no job was truly unskilled: She had to master new terms, new tools, and new behaviors for each one. And she quickly discovered that each job was mentally and physically exhausting.

Each of Ehrenreich's wage jobs presented a self-contained work culture, with distinctive personalities, customs, standards, and hierarchies. Some things never changed. Regarding standards, you shouldn't be "so fast and thorough you end up making things tougher for everyone else." And you learn not to "reveal one's full abilities to management, because the more they think you can do, the more they'll use and abuse you."

The U.S. culture of extreme inequality is revealed. "Corporate decision makers. . . occupy an economic position miles above that of the underpaid people whose labor they depend on," she writes. Her book documents many problems meeting life's needs, especially eating and sleeping. From her detailed cost analysis, Ehrenreich concludes "wages are too low and rent is too high."[8]

Nickel and Dimed gave me a glimpse into management systems from the

People are sometimes penalized for doing more.

perspective of the minimum-wage earner. I never realized how humiliating a routine drug test can feel. Or how demeaning some managers or supervisors are when conducting an intrusive, pre-employment interview or personality test. And the author discusses the psychological toll resulting from treating employees as untrustworthy. Top-down rules and regulations are designed to catch a potential slacker, drug addict, or thief.

In her words, "If you're made to feel unworthy enough, you may come to think that what you're paid is what you are actually worth."[9]

A common perspective

This is obviously a selective and biased view of management, but my numerous conversations with wage workers during more than 30 years tells me this perspective is not unusual. Most supervisors mean well, but many don't take time to understand the perceptions of those they manage. Instead, they enforce top-down generic rules that seem to consider wage workers as objects or a means to an end — rather than people with special needs, aspirations, emotions, and challenges.

The result: unhappy wage workers who don't contribute to production and safety as much as they could. That's unfortunate. Ehrenreich concludes from watching waitresses, retail workers, and housecleaners that "left to themselves, they devised systems of cooperation and work sharing…in fact, it was often hard to see what the function of management was, other than to exact obeisance."[10]

It's unlikely these workers will go beyond the call of duty to actively care for safety. They can be expected to do only what's required, and no more. That's what happens when workers are managed in a way that lowers their self-esteem and disconnects them from the organization.

The Sins of Wages

The Sins of Wages by William Abernathy[11] defines a major reason — perhaps a "root cause" — for the deplorable circumstances depicted in *Nickel and Dimed*. Most workers are not paid for what they accomplish but only for the amount of time they put in. This motivates a "just-put-in-your-time" mentality.

Combine that thinking with the perception that management doesn't care about the individual worker, and naturally peer pressure builds to do only enough and no more. Actively caring or going beyond the call of duty is out of the question.

Abernathy refers to the typical hourly-wage system as "entitlement pay" that contributes to an "entitlement culture." Employees believe they are owed their pay regardless of personal or company performance.

What happens when you have this kind of disconnect between job performance and wage compensation? Companies set up an accountability system — often based on behavior-based threats rather than recognition. They are implemented by supervisors who have not received effective behavior-management training, and have little empathy for the distressful plight of the minimum-wage worker.

Safety leaders often find themselves in the middle here, working between low-wage employees and command-and-control supervisors. They are working in a culture where supervisors are ill-trained, time pressures great, and employees unmotivated to go beyond the call for safety.

We don't envy any of the players in this scenario. What do to? Any attempt to improve safety here must begin with empathy — a critical quality of leadership.

6 Case Study: Observations of Yank and Aussie Cultures

I have had the pleasure of visiting Australia several times, and these trips have allowed me the opportunity to focus on some interpersonal differences between Americans and Australians. My perspective has been enhanced by a 1991 book by George W. Renwick, *A Fair Go For All: Australian/American Interactions*.[12] This book was given to me by an Aussie colleague to help me understand culture distinctions between the U.S. and Australia, and improve my presentations Down Under.

These are impressions from one scholar, often supported by my own observations. They reflect sweeping generalizations with room for many individual exceptions. Still, it is intriguing and perhaps thought-provoking to consider these cultural factors as potential determinants of social influence and the impact of a People-Based Safety™ process.

In fact, I'm more convinced the interpersonal factors discussed here influence the impact of PBS than I am that they validly differentiate the average American from the average Australian.

Developing relationships

Social psychologists claim it is human nature to want others to like us, and we are attracted by interpersonal similarities.[13] The first part of this sentence is probably true, but not necessarily the second part. While Americans are more likely to be interested in people who agree with them, Renwick claims Australians like disagreement. While Americans often view disagreement as rejection, Australians do not.

Renwick believes Americans determine interpersonal attraction rather quickly, whereas Australians are not so readily influenced by initial impressions. They take time to evaluate a person's character. Americans develop favorable perceptions from immediately available variables such as one's education, social status, material wealth, and recent achievements.

When people judge liking on the basis of surface phenomena, their interpersonal relationships change expeditiously. Renwick assumes Americans do this more often than Australians. He claims Australians take a longer time to establish friendship. Such relationships are long-lasting and meaningful, and include a strong sense of obligation. This relates directly to the principle of reciprocity.[14]

Showing gratitude

Thirty years ago, reciprocity was identified as a social norm. Behavioral research led social psychologists to presume people feel a sense of obligation to return personal favors — even if the recipient of the returned favor was not the source of the initial favor. This social-influence principle implies people who actively care for the safety and health of others motivate more actively caring between people. This can lead to the potential development of an actively-caring work culture.

But more recently our behavioral research suggests reciprocity may be waning in the U.S. For example, in one study we gave students $10 and 10 "actively caring thank-you cards" to use to recognize the desirable behavior of others.[15] But these students did not deliver more cards than a control group that only got the 10 cards. When we rewarded students with $1 per card delivery we indeed increased the number of "thank-you's" distributed campuswide.

Americans commonly say "no problem" after receiving a "thank-you" for doing someone a favor. This stifles the reciprocity cycle. They would support reciprocity if they said instead, "You're welcome, but you'd do the same for me."

Indeed, I must admit a reluctance to play the "reciprocity card." I say to myself, "I don't want to be obligated." Is this perceptual bias influenced by an individualistic, "I'll-do-it-myself" culture?

Seeing the big picture

Collectivism is analogous to systems thinking. It implies mutual interpersonal ownership of a problem or solution, as opposed to individualism, which gives precedence to individual initiative and choice over interests of the group. Policies in Australia, from gun control to traffic monitoring, imply a collectivistic perspective. The individualistic viewpoint is manifested in America.

Laying down the law

Renwick asserts Australians see themselves as inner-directed more than other-directed. So they are less apt to be rule-governed than Americans. For example, the country-wide doctrine for industrial safety in Australia is "duty of care." This reflects an overarching need to actively care for safety and health. But specific rules for doing this are notably absent in Australia, in stark contrast to the plethora of safety rules, regulations, and standards in the U.S.

Americans react vociferously against any attempt to intrude on their privacy, even for a societal benefit; yet are content with the imposition of lists of laws to follow. In contrast, Australians do not consider themselves controlled by outside directives, but rather by their inside character to do the right thing. This suggests cultural differences with regard to the principle of authority.[16]

Respecting authority

Social psychologists have shown people comply with the mandates of others in authority positions, even when the command is counter to sound judgment or common sense. While our surveys show that neither Americans nor Australians admit to being influenced by this authority principle, my personal observations suggest Aussies are more verbally resistant to authority control than Americans.

Many factors influence respect for authority.

George Renwick agrees, and claims Americans respect authority, whereas "Australians tend to denigrate authority because it acts as an external guide to making decisions and taking actions, and Australians are not comfortable with external controls."[17]

Accepting feedback

Given the cultural differences reviewed here, the acceptance of corrective feedback regarding one's behavior should be greater among Australians than Americans.

Why? First, Australians like disagreement, and unlike Americans, do not connect personal rejection with controversy or a difference of opinion. More importantly, Aussies do not judge people by their actions, but place more value on inner qualities. In contrast, Americans consider accomplishments from behavior a key measure of a person's merit.

Thus, it is easier for Australians to separate the outward and inward characteristics of an individual, and to deliver and accept corrective feedback as only behavioral advice independent of one's character or self-worth.

This final distinction between Australians and Americans is the most relevant for People-Based Safety™ (PBS). The prior cultural differences set the stage for this differentiation, which was not specifically studied in our survey research nor mentioned in Renwick's book. At any rate, since the success of behavior-based coaching is contingent on the delivery and acceptance of corrective feedback, coaching, a critical component of PBS (recall the ACTS model — Acting, Coaching, Thinking, Seeing) should go over better in Australia than in the U.S.

Bottom line

It is certainly risky to conclude from a few sweeping generalizations about two cultures that corrective feedback, a critical component of PBS, will work better in Australia than the U.S. I'd have substantial confidence in this prediction if these cultural distinctions were valid and reflected characteristics of a majority of the individuals targeted with a PBS process. In other words, the interpersonal factors in this discussion of cultural differences are more likely related to PBS impact than the culture presumed to reflect them.

To enrich your culture, remember: Individuals are more likely to accept behavior-based corrective feedback and react appropriately when they a) accept and appreciate disagreement; b) base relationships more on inner character than observable behavior; c) accept and honor obligations to return favors; d) adopt a collectivistic mindset regarding industrial safety; and e) see themselves as more inner-directed than other-directed.

7 Case Study: Inspiration to Actively Care

For more than two decades I've decried the use of the word "accident" within the context of industrial safety. When something occurs "accidentally," chance and uncontrollable factors are implied. This is not the case for most workplace injuries. Usually someone in the system — either a manager, a coworker, or the victim — had the knowledge and resources to prevent the mishap, but made a risky decision.

But here, however, I am reflecting on a true accident. I'm referring to amyotrophic lateral sclerosis (ALS), better known as "Lou Gehrig's disease." This devastating illness attacks the bodies of more than 5,000 people annually, and ends a person's life in two to five years. The disease causes successive loss of voluntary muscle contraction, eventually resulting in complete muscle paralysis, inability to communicate, impaired swallowing and breathing, and ultimately death. Mitch Albom, author of *Tuesdays with Morrie*, summarizes ALS as follows:

"ALS is like a lit candle: it melts your nerves and leaves your body a pile of wax. Often, it begins with the legs and works its way up. You lose control of your thigh muscles, so you cannot sit up straight. By the end, if you are still alive, you are breathing through a tube in a hole in your throat, while your soul, perfectly awake, is imprisoned inside a limp husk, perhaps able to blink, or cluck a tongue, like something from a science fiction movie, the man frozen inside his flesh."[18]

No choice

Dick Sanderson grew up in Mendham, N.J., and now resides in Charlotte, N.C. He was diagnosed with ALS a little over a year ago, and the progression of the disease has been rapid. Once a dedicated and talented athlete, Dick can now only move his head slightly and show some facial expression. He receives comprehensible input from the world around him, but can offer only minimal output.

Dick can smile, and to my surprise he smiled more than anyone else when I met him two years ago. The steadfast courage Dick Sanderson and his family have shown throughout this onerous and emotional ordeal is inspirational.

For me, the saddest part of this story is that no one did anything to cause this heartbreaking situation. Yet no one complains. Dick's wife, Dawn,

Social support helps us overcome overwhelming challenges.
Pictured (from left to right): Joanne Dean, Dick Sanderson, Scott Geller

cares for him intently and continuously, with periodic assistance from their

16-year-old son, Trey, and college-aged daughters, Amy and Melissa. On a weekend visit, I saw Dawn feed Dick through a stomach tube, and witnessed friends and family suction saliva from Dick's mouth several times per hour.

Seeing this level of actively caring is instantly humbling and sobering. And throughout all of this, Dick Sanderson displays incessant gallantry and grace, as well as a sense of humor. For example, while one of Dick's high school friends was singing a song written as a tribute to the Sanderson family, Dick signaled for suctioning as if to say "stop singing." The audience appreciated the humor of Dick's request and laughed aloud.

Support of friends

Besides a loving family, Dick has a large circle of dedicated friends

Dick Sanderson with his 1969 classmates of West Morris Regional High School, Chester, New Jersey.

supporting him and his positive fight with ALS. When Dick Sanderson revealed his ALS diagnosis, people came from near and far to give personal testimony to his limitless influence in their lives. Numerous friends from his high school days 37 years ago came, as well as current basketball buddies, business partners, parents of children Dick coached, and friends of their friends.

Psychological research has demonstrated the unique value of social support during times of personal struggle — both emotional and physical. But friendship is more than social support, and Dick Sanderson has been an exemplar for developing and cultivating true friendship. Even with all his athletic and artistic talents, Dick has always been humble and friendly to everyone, regardless of their stature and abilities.

Living in the moment

When asked how she does all she needs to do to care for her husband, Dawn replies, "I'm just loving him." She writes, "This is a very terrible thing we have to face, but knowing we are not facing it alone is so powerful. . . . We don't look forward or focus on the future, but rather we look back and feel the blessings of the rich and fulfilling life we have shared together."

In a similar vein, Rod Dorman, a longtime friend of Dick who also has a terminal disease, writes, "It is not about how many days on earth you live; it is about how you live each day. 'He who lives the most each day wins'. . . . All of us have commented 'How time flies'. . . . That is no longer true for me. Confronting a terminal illness slows time. Every day, I reflect on the fact I have that day to live, and I evaluate whether I am living it well."[19]

Relevance to safety

Appreciating the moments of our lives enriches a culture, be it a family or organizational culture, in two ways.

First, realizing the good fortune of a healthy life motivates us to preserve this gift, including the performance of those somewhat inconvenient safety-related behaviors.

Second, when we are mindful of the present moments of our daily lives, we are less likely to make errors that can put ourselves and others at risk for injury. For example, many multitasking activities, like using a cell phone while driving, put people at risk for unintentional injury, and are clearly contrary to present-moment thinking and living.

Multitasking puts people at risk for injury.

We need each other

Sometimes we multitask because we feel overwhelmed and don't have enough support from others. But do you ask for help, or do you feel a need to be independent? Perhaps you perceive you haven't cultivated a level of friendship with coworkers that warrants a request for their help.

But you can't do it alone. Fortunately, you are not as dependent on support from others as is Dick Sanderson. But you cannot be as safe as you can be without assistance from friends and colleagues. Sometimes you need help to complete a job safely. Or you may need advice on how to eliminate a hazard. And everyone can benefit from the behavior-based feedback of an actively-caring coach.

No one can improve their competence without behavioral feedback, and when it comes to safety, such feedback must come from an observer. Of course, this level of interdependency requires the right work culture — one with people who realize they need each other to help them and their organization be the best they can be, while remaining injury-free.

The power of choice

Why should you contribute daily to enriching a trusting and interdependent work culture? Because you can. You have the power of choice. You can choose to be mindful of everyday hazards and act accordingly, and you can choose to actively care for the safety of others. You can choose to do whatever it takes to maintain your current skills, abilities, and physical condition. Dick Sanderson does not have that choice.

So much to lose

Dick Sanderson's friends convened the weekend of my visit with a friend of his to show him love and admiration. Sadly, one by one they also said "Goodbye." I will never forget this bittersweet weekend. I will use it as inspiration to appreciate what I have today and what I have to lose if I choose to take a risk at the wrong time.

Think of Dick Sanderson smiling through his enormous health challenges — an accident over which he had no control. Now, how can we complain about daily disappointments? How can we justify unsafe and risky acts? Instead, we can smile in knowing we have the power to choose safety over risk and avoid unthinkable loss.

CHAPTER FIVE

Improving Communication

Introduction

If you want to learn about a culture, listen to what is going on around you. Pay attention to the conversations around the dinner table, in the locker room, on a work break, in a cafeteria or hallways. Conversations, how people talk to each other and what they talk about, are the living, pulsating heartbeat of a culture. They will tell you what a family, a team, a school or a business believes and values.

"I can't tout the benefits of conversation enough," says Dr. Geller at the beginning of Part 1. He proceeds to show leaders five ways they can use conversations to enrich their cultures, and their personal relationships. "When it comes to injury prevention (and culture enrichment) we can't have too many quality interpersonal conversations," he concludes.

One sure-fire way to stifle conversation is to come off as a know-it-all. Part 2 makes the point you can't lead people without earning their trust. So arrogant leaders need not apply. Remember, says Dr. Geller, people trust leaders who are consistent, communicative, caring, candid, committed, oriented to consensus, and invested in character. These constitute the "Seven Cs" of trust-building. If you want to enrich your culture, practice the Seven Cs.

Parts 3 and 4 make the case that leaders need to ask more questions. How can you sustain any conversation without someone doing the probing and listening? The same holds true for a culture. You can't make it richer, filled with motivated and committed people, unless someone is sustaining, and subtly shaping in a positive direction, those powerful conversations that constitute a culture. That's a leader's role.

Ask more questions, says Dr. Geller, and you show you value the input of others, raise their self-esteem, receive important feedback, and improve your own competence.

Remember, leaders do not bail out in the face of possible failure. They don't hold back their requests for support because they fear rejection. One of the toughest challenges in safety is to ask for support, either from busy supervisors, inaccessible executives, or "What's in it for me?" employees.

Sometimes, as Dr. Geller explains, the non-directive approach works. To gain buy-in, drop your agenda and ask about barriers — why supers aren't spending more time on safety, or why employees aren't wearing PPE. Then ask what you can do to help get around those barriers.

Other times, you must be direct and to the point. Keys to this type of communication: keep your request simple, resist the urge to argue when the response you get is disappointing, and be politely persistent. Keep the door open for re-asking, and have a sense of timing for when it's best to come knocking again.

In Part 5 Dr. Geller advocates judicious use of an audacious form of communication — laying on a guilt trip. The term "guilt" carries sufficient negative baggage for most leaders never to use it. They feel it threatens trust and their credibility. But Dr. Geller shows how leaders can finesse an inner tension between an individual's pro-safety values and their at-risk behavior. Individuals will likely put an end to that irritating tension by choosing to live up to their stated values and perform safely in the future.

The good professor knows of which he speaks in Part 6. Living in both the academic and consulting worlds (or cultures), he shows how better communication can bridge the differences between these diverse communities. It is a reminder of how conversations and other forms of communication are indeed important bridges, connections that help bind cultures and organizations together. Effective leaders know how to build bridges.

Dave Johnson, Editor
Industrial Safety & Hygiene News

① Powerful Conversations

As a safety leader, what part do conversations play in your safety improvement process?

I bet most, if not all, of your attempts to improve workplace safety include interpersonal communication. In the People-Based Safety™ model ACTS, as applied in this book to leaders, the "C" stands for leadership coaching. And a key component of any coaching effort involves interpersonal conversation. The ability to engage coworkers in conversation is critical to being an effective leader. And an organization where engaged conversations occur at all levels is enriching its culture.

I can't tout the benefits of conversation enough. It is a key component of behavior-based observation and feedback, peer-to-peer coaching, employee recognition, incident analysis, and corrective action development and implementation. Engineering benefits from learning human dynamics, too, and this requires an interpersonal exchange of ideas and perceptions.[1]

The success of any intervention involving people depends on communication. In this article I define five types of safety-related communication, each playing a particular and essential role in safety-related intervention. I heard these conversation labels during a Progressive Business audio conference in 2003 featuring Bob Aquadro and Bob Allbright. I think safety leaders will find it useful to consider how interpersonal conversation varies in these five ways.

1 — Relationship conversation

Simply put, these conversations occur whenever you show sincere interest in another person, from their home life to their work challenges. This happens, of course, when you talk about particular aspects of a person's family, health, hobbies, work processes, or safety-related perceptions.

As Dale Carnegie said years ago, and echoed later by Ken Blanchard and Spencer Johnson, "Help people feel important at doing worthwhile work."[2] This is relationship-building.

Specific behaviors you find desirable might surface during a relationship-building conversation. If so, certainly acknowledge their occurrence and show appreciation. But your intention is more about developing support and interpersonal trust than influencing behavior. You want to remove any perception of manipulation or behavior modification. The key is to show genuine interest in the other person's situation, performance, or perspective.

2 — Possibility conversation

These conversations occur when you share visions with another person. Of course, a prime industrial-safety vision is "injury-free." But possibility conversations target any future situation that reflects desired improvement in environment/engineering conditions, your organization's culture, individual

Negative reinforcement feels like manipulation.

behavioral competence, or person states.

In his 1977 book, *The Art of Leadership Conversation*,[3] Kim Krisco recommends we begin coaching conversations with a discussion of a person's past, analogous to the relationship conversation discussed above, then progress to a discussion of future possibilities, as defined here.

Krisco proposes the coaching conversation transition back to the present, so people can define process goals or behavioral strategies relevant to achieving certain possibilities. The next three types of conversation reviewed here are actually subcategories of this suggestion by Krisco for change-focused interpersonal coaching.

3 — Action conversation

Here we have behavior-based communication. Given a vision or possibility for improvement, this conversation focuses on what an individual or work team could do to move in a desirable direction. The conversation might be between individuals, as in coaching, or between members of a group.

The action conversation could define a number of different behaviors, some to continue and others to decrease or eliminate. When these conversations occur in group meetings, individual assignments are often needed. Also, action goals are set according to the SMART acronym (for Specific, Motivational, Achievable, Relevant, and Trackable).[4]

This goal-setting exercise should include an accountability system for tracking progress toward goal attainment. With work groups or teams, it's usually best to monitor not only individual achievements with regard to specific assignments, but also the group's progress as a team.

4 — Opportunity conversation

OK, you've learned how to do behavior-based observation and feedback, and set a goal for completing a certain number of observation/feedback sessions in one month. Now it's time to look for opportunities to conduct such a one-on-one session.

In some cultures, this can be any work situation that involves human behavior. But in other industrial settings, workers must agree to be observed before the process can be implemented.

Suppose an individual or work team chooses to adopt an achievement or success-seeking perspective to safety by tracking all safety-related behaviors performed beyond a person's daily work routine. This requires an action conversation about the types of behaviors that indicate "going beyond the call of duty," and an opportunity conversation about the various situations that call for designated safety-improvement behaviors.

Bottom line: A practical action plan for achieving particular possibilities includes a definition of behaviors and situations — behaviors needed to fulfill the plan (an action conversation) and the times and places for these behaviors to occur (an opportunity conversation).

5 — Follow-up conversation

It's important to acknowledge the achievement of a SMART goal. These follow-up conversations are rewarding, and promote a success-seeking mindset. After noting the acquisition of an action/opportunity outcome, a follow-up conversation turns to discussion of a subsequent challenge. This could include conversations 2, 3, and 4 — an identification of new possibilities (2), relevant and acceptable action plans (3), and opportunities calling for certain action (4).

Follow-up conversations target the end result or outcome of an action plan but they often focus on the process first.

Suppose, for example, you communicate with a supervisor regarding a need to have more one-on-one interaction with line workers. After exploring possibilities, you discuss specific actions and opportunities for meaningful supervisor/employee contacts. You might set a SMART goal and even a follow-up reward for goal attain-

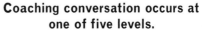

**Coaching conversation occurs at
one of five levels.**

ment. But process-focused monitoring could also be quite helpful. In other words, it would probably be useful to contact this supervisor periodically for follow-up conversations regarding his or her progress toward goal attainment.

Please note: One type of conversation does not stop with the implementation of the next in the sequence. Relationship conversations, for example, continue throughout action planning, accomplishment, and follow-up. And while it makes sense to define the behaviors in an action plan before considering opportunities, in actual practice people look for opportunities for their action-plan behavior before performing.

Indeed, interpersonal communication will fluctuate between all five conversation types. Perhaps understanding these different conversations and their objectives will help increase the quantity and improve the quality of your diverse communications as a safety leader. When it comes to injury prevention, we can't have too many quality interpersonal conversations.

10 Keys to Conversing

1) Display sincere interest in the person, or group, you are talking to.
2) Develop the other person's trust, or the team's trust, in you.
3) Remove any perception of manipulation or cold behavior modification.
4) Share your vision with the other person or group regarding safety in your organization.
5) Shift the conversation's focus to what that person, or team, can do to achieve your goal.
6) Make individual assignments for reaching the goal.
7) Set up an accountability system to track progress toward the goal.
8) Throughout your conversations about your goals, always specify the required behaviors to reach the goal.
9) Always follow up to talk to individuals about their progress, and to recognize their goal achievements.
10) Remember, one type of conversation does not stop when the next begins. Communication continually weaves together all five types of conversation studied here.

2 "Know-It-Alls" Need Not Apply

Some business managers are stuck in the past. They believe we are still in the Industrial Age of top-down control. They do not appreciate nor embrace a paradigm shift in the 21st century, identified by Peter Drucker as the "Age of the Knowledge Worker."[5]

Consider the "Blue-Collar Band of Brothers," described by Joanne Dean in her *Industrial Safety & Hygiene News* article.[6] This "band" of construction workers contributes more than manual labor. They bring unique skills and experiences to their jobs, and willingly offer thoughtful and relevant advice when asked. They are open to learning and applying new ways to making work routines more effective and/or more safe.

But will their knowledge be used to improve job safety?

Will they receive opportunities to learn and apply the human dynamics of injury prevention?

Leaders from both management and union sides of an organization determine whether people-based skills are learned and used to increase safe production.

What kind of leader is open to receiving advice from hourly workers regarding ways to achieve more safe production?

What does it take for employees to provide input for decisions relevant to the safety of their jobs?

Leaders tap the knowledge of hourly workers. Pictured (from left to right): Jeff Lucarella, Bob Taylor, Richie Schoene, Joanne Dean, Richie Meyler, Kevin Worthington.

How can leaders encourage knowledge workers to hold their peers accountable for safe work practices and promote the development of self-accountability?

Leading the knowledge worker

In his latest book, *The 8th Habit: From Effectiveness to Greatness,*[7] Stephen R. Covey provides a concise answer to these questions. Dr. Covey advises leaders to communicate the right vision and set high-priority goals, while constantly looking for the potential in people by "modeling the courage to determine a course and the humility and mutual respect to involve others in deciding what matters most."[8]

For safety, the Covey paradigm implies doing more than holding people accountable for top-down safety rules. While OSHA rules define a general course of action for injury prevention, and certainly have their place, leaders need to empower workers to provide their own action plans and accountability systems. This is more than rule-following behavior. It's an ongoing, interactive process of team members identifying hazards and potentials for personal injury and then defining specific ways to avoid them. This is the safety responsibility of the knowledge worker.

What kind of leader makes this happen?

7 Core Components of Trust [9]

1. **Communication** — exchange of information or opinion by speech, writing or signals
2. **Caring** — showing concern or interest about what happens
3. **Candor** — straightforwardness and frankness of expression; freedom from prejudice
4. **Consistency** — agreement among successive acts, ideas, or events
5. **Commitment** — being bound emotionally or intellectually to a course of action
6. **Consensus** — agreement in opinion, testimony or belief
7. **Character** — the combined moral or ethical structure of a person or group; integrity; fortitude

The humble leader

The Covey quote on page 101 includes the leadership quality I believe is the most critical for obtaining safety-relevant information from knowledge workers. Managers and safety professionals alike need to publicly acknowledge they don't know enough to keep their employees safe. Indeed, workers on the job are in the best position to observe potential for injury and recommend practical prevention strategies.

The humble leader continually asks advice of followers with relevant experience. Realizing behavior-based feedback is essential for improvement, the humble leader also asks for feedback. When a worker answers "OK" to the leader's question, "How am I doing?" the humble leader then asks, "What can I do differently to be better?"

With a quick-fix answer of "nothing," the humble leader probes for more candor. "Come on, no one is perfect. Tell me one thing I could do more or less often to promote safety."

Some might be reluctant to reveal an observed weakness in a leader. Why? Because they fear negative consequences. They have insufficient trust.

The trustworthy leader

Dr. Covey reminds us that "trust" is both a noun and a verb. In other words, you can have confidence in the integrity, truth, ability, intentions, and character of a person. But you also trust when you rely or depend on people to meet your expectations. Sometimes entrusting a person to go beyond the call of duty for safety motivates relevant action. Dr. Covey puts it this way: "Trust becomes a verb when you communicate to others their worth and potential so clearly that they are inspired to see it in themselves." (p. 181)[10]

I discuss ways to build trust in *People-Based Safety™: The Source.*[11] These trust-building interventions are organized around seven C-words: consistency, communication, caring, candor, commitment, consensus, and character.

These words and their dictionary definitions activate a variety of practical ways to augment the extent people trust the intentions and/or abilities of others, particularly their leaders. I recommend using the sidebar on page 102 to stimulate trust-building discussions in group meetings. Ask participants to define specific behaviors that reflect each of the trust words. You might also ask the group to describe occasions when a particular trust word was lacking, and then to suggest improvement strategies.

A trusting leader facilitates trustworthiness.

To conclude

These leadership recommendations are certainly appropriate for improvement domains beyond safety and for settings beyond the workplace. Bottom line: Continuous improvement depends on continuous input from people with relevant information, and leaders who are humble and trusting to receive such information.

Qualities of the Humble Leader

- Publicly acknowledge you don't know enough to keep your employees safe.
- Acknowledge people on the job are in the best position to observe potential for injury and recommend practical prevention strategies.
- Continually ask advice of followers who possess relevant experience.
- Ask for feedback on your own performance.
- When a worker merely answers "OK" to your question, "How am I doing?" next ask, "What can I do differently to be better?"
- If you receive a quick-fix answer of "nothing," probe for more candor. "Come on, no one is perfect, tell me one thing I could do more or less often to promote safety."
- Your probing also serves to build trust between you and the person(s) you're talking with.

③ Can I Ask You a Question?

"There's no substitute for knowledge."

Years ago, I heard W. Edwards Deming say this several times throughout a four-day workshop on Total Quality Management.[12] Now I repeat this same phrase in every university class I teach. It's a powerful principle and it's key to improvement, whether the focus is on safety, productivity, quality, or relationship-building.

But how do we obtain knowledge?

Knowledge comes in many forms. It's public or personal, it's objective or subjective, it's understood or misunderstood, it's useful or useless, it's considered or ignored, and so on. My purpose is not to explore various types of knowledge, but simply to consider how we gain knowledge for improving workplace safety. Bottom line: Leaders need to ask more questions.

Why ask?

Let's begin with the end in mind. Why should leaders ask more questions about safety? Here are five benefits:

1) To "Always Seek Knowledge"

Here's the most obvious reason for asking: "Always Seek Knowledge (ASK)." You learn about other people's behaviors, attitudes, feelings, and perceptions by asking them directly.

People often project their mistakes beyond themselves.

Sure, you usually have a personal opinion about why someone is taking a risk, ignoring procedures, and so on. But ask for the other person's perspective first. Even though the reaction might sound defensive, accept it as knowledge you need to completely comprehend the situation. Then, after showing genuine appreciation for the other person's outlook, you can expect that person to consider your interpretation.

2) To show you care

Our busy lives often prevent us from taking the time to ask more than, "How are you doing?" and listen to anything more than, "I'm OK, thanks for asking." We might take more time if we realized the powerful impact of asking.

For example, Joanne, a safety leader, was asked to give an orientation session to four new employees of the construction division of her company. She gave a brief overview of the company and reviewed standard safety rules and guidelines. Then she initiated a lively discussion by asking each employee their viewpoints about safety and their job expectations. The two-hour orientation expanded to a four-hour sharing of personal experiences.

Joanne told me she learned valuable information about the interests and talents of four different individuals, and she answered thoughtful questions from her audience. The knowledge gained from this interactive asking process enabled her to customize the safety information, making the material more meaningful and relevant.

3) To raise self-esteem

Joanne's interest in asking personal questions likely enhanced some of the new hires' self-esteem. Whenever you ask a person for their advice or a personal opinion, you pump up their sense of self-worth. This is especially true when the person doing the asking is a respected and credible leader. In this situation, for example, a person might think, "The experienced safety professional for this company is asking for my opinion. I must be important in her eyes."

4) To obtain feedback

Only with feedback can you improve your performance. Sometimes your behavior provides natural ongoing feedback — when you see the results of painting with a brush, writing with a pen, or hitting a golf ball. In most cases, though, you don't get enough natural feedback for optimal performance. Even artistic and written expression benefits greatly from critiques.

Much of the feedback needed for competence-building is extrinsic, often coming from an observer. For example, a behavioral coaching process can assure the safe performance of a whole work team through interpersonal observation and feedback.[13]

Leaders ask relevant others for feedback.

Leaders who want to develop a continuous improvement mindset throughout their work culture should periodically ask for feedback about their own performance. The more interpersonal feedback requested and given throughout a workplace, the more performance improvement is possible, whether for safety, productivity, quality, or relationship-building.

5) To improve personal impact

By asking for feedback you improve your competence, and you stand to make a bigger positive difference in your role as a safety and health leader. You can also enhance your impact by asking for more support. Do you ever resist asking for certain support because you assume the worst?

My friend, a safety leader for a successful company, is extremely passionate about learning as much as possible about the prevention of occupational injuries. I periodically inform him about upcoming professional development conferences that provide outstanding opportunities to add useful strategies to his "safety tool box." The reaction I usually get is, "Oh, my company won't allow me to count conference attendance as work time, and they certainly won't cover my expenses."

Maybe a request for support is a long shot, but why not ask anyway? Many of us hold back our requests for support because we fear rejection. We ruminate to ourselves, "Why waste time on a lost cause?" My next essay in this chapter shows you how to go about soliciting support.

Only by Asking...

- Do you get the other person's perspective.
- Do you learn something about that person's values, beliefs, experiences, and expectations.
- Do you build self-esteem in the person you're conversing with, enhancing their potential to be a contributor.
- Do you obtain information to improve your own performance and competence as a leader.

4 If at First You Don't Succeed...

Most of us don't ask enough questions in the course of our conversations. Yet, asking is the key to learning. By asking for feedback we strengthen relationships by building trust. By asking for someone's opinion, we show caring and respect. This sets the stage for frank communication. By asking for advice we gain information and boost the other person's self-esteem. And by asking for support, we increase the chances of actually getting it.

But how should we ask? Leaders, who often find themselves in the position of requesting resources or other forms of support either from managers, supervisors, or employees, can take a direct or non-direct approach, depending on the situation.

No agenda

What do I mean by non-directive asking? You could say this is "asking without an agenda." Counselors use the non-directive approach to draw out their clients.[14] They carefully avoid personal judgment or interpretation while listening patiently to their client's stories. They respect the distinct views of every individual, and they don't make comparisons or generalizations between the stories of different people.

This line of questioning can be quite useful in obtaining information relevant to developing a safety process — while gaining buy-in at the same time.

Here's how it can work: You ask a coworker for his or her opinion on the best way to set up a safety procedure. You're being genuine, with no hidden agenda. A frank discussion about possible safety guidelines

"WHY AREN'T YOU WEARING YOUR GLOVES?"

Confrontational conversation does more harm than good.

follows. This leads to a customized protocol. And workers will "buy in" because they had an opportunity to offer input.

Here's another example: Say you observe an at-risk behavior. Telling someone they are not following safety guidelines can sound insulting and put people on the defensive. Will they feel responsible and self-motivated for the long run? I doubt it.

Taking the non-directive approach, you could note that certain personal protective equipment is not being used, and ask, "In your opinion, why is that PPE unpopular?" Or, "What can I do to facilitate the use of that PPE?"

Assume there are legitimate barriers to the safe behavior you want to see, and there are ways to remove at least some of these barriers. Who knows better how to address this problem than the workers themselves?

Also, believe that most of the workforce wants to help prevent personal injury. With these reasonable assumptions, non-directive asking seems to be a most sensible way to discuss an at-risk behavior.

To the point

Sometimes leaders need to be more direct in asking for something. Leaders might need to ask for specific safety resources, personnel assistance, or opportunities for professional development.

But many of us would rather not ask for help, or whatever it is we need. Maybe we think it's a sign of weakness, or stupidity. Maybe we fear rejection. These are irrational excuses, really, nothing more. Asking is key to acquiring knowledge and building competence. People who ask for feedback or support are trying to be the best they can be.

The second excuse for avoiding asking reminds me of a story I heard about Dr. Albert Ellis, the renowned clinical psychologist who developed the popular and effective cognitive therapy entitled Rational Emotive Behavior Therapy (REBT).[15] Specifically, Dr. Ellis has reported that as a college student he never had difficulty getting dates, unlike his more handsome friends. Why? Because he simply asked many more girls for a date than did his friends. Why? Because he didn't fear rejection.

You see, Dr. Ellis did not entertain the global notion that "Everyone must like me." Rejection is normal and to be expected. So keep asking for what you want, and eventually your request will be honored. Of course, you can help get what you want by knowing how to ask.

Follow these suggestions:
Drop the act of over-competence

No one is perfect, and everyone can improve. So whether you're asking for feedback or support, show a willingness to be vulnerable. Sure, you're good at what you do. But you could do more with the kind of support you're asking for. Speak from your heart with genuine desire to make a bigger difference.

When you ask for support, be ready to answer the question, "What's in it for me?"

People are more likely to honor your request when they see possible benefits coming their way.

You might not be able to estimate the ROI (or return-on-investment), but you can certainly explore possibilities. What long-term consequence could be gained from an investment in certain support?

Also, realize people are motivated by more than money. As a teacher, for example, I'm persuaded to offer extra support for students when I believe the effort will enable them to become more competent contributors to humanity.

Don't be impulsive

Consider the costs and benefits for the support you need. Be confident the positives outweigh the negatives. Your challenge is to sell this view to the person whose support you need. So while you're humble, you're also convinced the support is advantageous from both a personal and organizational perspective.

Keep your request simple

Be direct. Time is precious for all of us, of course. Specify what you want and when. Then, explain clearly why you want it, in terms of both short-term and long-term positive consequences.

List the benefits as definitively as possible. Then listen attentively and patiently for a response. Wait for a complete reaction before mounting a defense, if needed. Usually, you'll only get questions, which you can answer precisely and concisely because you anticipated them and prepared answers beforehand.

When the reaction is disappointing, resist the urge to argue

That is, unless you are making a last-resort request and the support is necessarily "now or never." But even under these circumstances, you need to sustain a congenial atmosphere if you have any hope of reversing the decision.

If the support you request can be delayed, ask for an opportunity to revisit your request at a later date. You're trying to get around a flat-out "no" to your request. Remember, the more often a person says "no" to a request, the more difficult it is for this person to say "yes" at some later point. This is especially true if your exchange becomes confrontational. It's important, but not easy, to remain as friendly and cordial as possible in the face of a disheartening decision.

If at first you don't succeed in getting support, assess what happened

Perhaps you became aware of costs you hadn't anticipated and benefits you exaggerated, at least in the eyes of the person who must honor your request. So now you have an opportunity to regroup and ask again. At least you remained good-natured and avoided conflict.

Keep the door open for re-asking, and persist when you know you're right. Timing is often critical. On another day, you'll likely make a more powerful request, especially if you consider each of the recommendations offered here.

5 The Power of a "Guilt Trip"

I bet most of you have used the term "guilt trip" when explaining personal feelings or when attempting to understand the behavior of others. What do we mean? Can leaders use this metaphor to improve safety?

Yes.

Put aside the negative connotations of "guilt trip." I think leaders can put people (including themselves) on a beneficial guilt trip for safety. Personal responsibility for injury prevention will be increased in the process. The idea is not as repugnant as it might seem. Let's explore this notion and its practical applications and ramifications.

What is guilt?

The behavioral definition of guilt is quite simply being responsible for an offense or "wrongdoing," as revealed in the first definition of guilt in *The American Heritage Dictionary*. Then there is the related legal definition of guilt — "culpability for a crime or lesser breach of regulations".[16]

But the term "guilt trip" reflects more than behavior. It suggests a feeling state. Presumably it motivates certain subsequent behavior. This is reflected in the fourth definition of guilt in my dictionary — "remorseful awareness of having done something wrong." Of course, the key word here is "remorseful" — which means one feels "moral anguish arising from repentance for past misdeeds".[17]

Psychologists call such mental anguish cognitive dissonance.[18] It's typically caused by an inconsistency realized between one's actions and a related belief, value, or attitude. This inconsistency between inside conviction and outside behavior causes mental tension and self-directed motivation to restore congruity between behavior and one's belief, value, or attitude.[19]

Simply put, we want our actions to reflect our values, and vice versa. When we perceive an inconsistency between our actions and our values, we more often adjust our behavior to match our values than change our values.

When do we feel guilty?

According to these definitions, a guilt trip motivates personal responsibility to ease inner tension and conflict and do something to reduce a perceived discrepancy between personal conviction and behavior.

When circumstances prevent me from getting proper sleep, nutrition, or exercise, I feel guilty until my actions restore balance. Likewise, when I don't give appropriate attention to family matters, I feel guilt or tension until my behavior reflects the appropriate attention to family values has been restored.

But what is "appropriate"?

Ah, there's the rub. It's left to each of us to decide how much action is needed to relieve our guilt and restore a sense of balance or consistency between internal conviction and external deeds. How to eliminate guilt through action is personal, subjective, and quite variable among individuals. Leaders need to keep this in mind.

Can leaders influence how much guilt a person feels and/or the quality and quantity of action needed to stop a guilt trip? My "yes" answer to this question has critical implications for increasing people's self-accountability for safety. In particular, if leaders can make a person feel guilty about performing a certain at-risk behavior, they will increase the likelihood that individual will accept personal responsibility for behavior change. Leaders might even activate safety-related action beyond correcting one behavior. To eradicate guilty feelings, a person might feel he or she needs to do more for safety than adjust one behavior.

Does this sound far-fetched? I'll discuss some practical ways to make this happen, but first we need to consider a popular slogan among safety leaders, "Safety is a value."

Is safety really a value?

Many years ago, in a 1991 column for *Industrial Safety & Hygiene News* magazine, I proposed safety should be considered a "value" rather than a "priority," and offered a reasonable rationale for this.[20] Since then, I've heard support for this opinion from numerous safety leaders and consultants.[21]

When I ask my audiences these days whether they hold safety as a value, almost everyone raises a hand to affirm a collective "yes." And when I question individuals, I invariably receive an assertive confirmation. Some say, "Safety is more than a value to me and my work team, it's a *core* value."

Frankly, I believe many people are too cavalier and quick to affirm safety as a core value or personal principle that receives precedence in every situation. But I'll accept their thinking, because this is the first step toward putting people on a guilt trip and helping them develop self-accountability for safety.

"OK," I say, "Safety is a value linked to every one of your priorities. So regardless of the circumstances, including outside demands on your time, you attempt to be as safe as you can be. Right?"

Surprisingly, most people answer "yes" to this idealistic proposal, especially in public settings. Actually, I shouldn't be surprised by this response. "Safety" goes along with "motherhood, baseball, and apple pie." There exists a certain peer pressure to be supportive of safety in a group context, or even when confronted one-to-one.

Note what I'm doing here.
• First, I get agreement that "Safety is a value."
• Then I explain the true meaning of this statement.
• To be consistent, listeners give successive "yes" responses, up to the final impractical affirmation that safety gets preference over everything. This defines safety as a person's "Number One Value." This confirmation certainly can be used to activate a guilt trip and build self-accountability for safety.

From value affirmation to responsibility

This intervention process is simple and straightforward. Get people to declare safety as a value linked to the changing priorities of each workday. Then define behaviors compatible versus incompatible with this value statement. A guilt trip can be activated whenever you point out behavior inconsistent with safety as a core value.

After observing an at-risk behavior you might remind the performer of the group consensus that safety is a value. If the person realizes the inconsistency, he or she should feel guilty and proceed to resolve the tension or attitude/behavior imbalance by substituting safe for at-risk behavior.

But if you want self-accountability, you need to enable choice in this situation. The individual's conviction should not be viewed as controlled by extrinsic contingencies like incentives or peer pressure or someone putting a guilt trip on them, but by the personal decision to demonstrate safety as a core value.

I describe the "hypocrisy effect" as a way to motivate behavior by having participants experience a discrepancy between what they have done and what they should do in *People-Based Safety™: The Source*.[22]

This intervention procedure is as follows:

1) Present the rationale for certain safe behavior;

2) Ask participants to commit to always choosing this behavior over designated at-risk practices; and

3) Ask participants to list the most recent times they performed at-risk alternatives to the safe behavior.

Do you see how the hypocrisy-based intervention mimics the guilt-trip tactic? The goal is to enable the target individuals to experience a discrepancy between their behavior and a commitment, or personal value statement. This initiates a guilt trip, relieved when self-directed action restores the balance.

The more public the commitment or value affirmation, the greater its impact. When people attest to safety as a value in the presence of others, they feel a greater sense of obligation or duty to live up to their avowal.[23] And those who heard the value statement can readily start that individual on a guilt trip by calling attention to at-risk behavior they observe that does not reflect safety as a value.

Use this value declaration to increase self-accountability for preventing injury — and as a means of enriching your Total Safety Culture. Simply start people on a guilt trip by revealing an inconsistency between their internal conviction that safety is a value and their at-risk behavior. As a leader, you must reveal this contradiction with empathy and signs of genuine caring, not as a "gotcha" ambush sprung by a "sniggling" superior.

Individuals will likely end their guilt trip by choosing the safe alternative in the future. They might even go beyond the call of duty for injury prevention in order to reaffirm safety is, indeed, a core value for them.

Leading with the "Hypocrisy Effect"

- Present the rationale for certain safe behavior;
- Ask participants to commit to always choosing this behavior over designated at-risk practices; and
- Ask participants to list the most recent times they performed at-risk alternatives to the safe behavior.

6 Can We Talk? Academics vs. Consultants

Is it possible to achieve synergy, agreement between two disparate groups, through conversation? Leaders sometimes find themselves in the position of trying to bring opposing groups together. I'd like to explore the distinct differences of two groups I'm well familiar with, and the possibilities of reaching a positive synergy between the two.

Comparing academics and consultants

Five years ago a major consulting firm sent its clients an email, in an attempt to dissuade them from attending the annual Behavior Safety Now (BSN) conference. The email stated disappointment in the quality of the conference presentations, and added, "The conference has become a showcase for academics and 'wannabe' consultants." That year I was a keynote speaker at the BSN conference, as I have been every year since its inception.

I began my address with a display of select portions from this email, including the quote given above.

Then I admitted to being both an academic and a "wannabe" consultant.

Yes, I'm proud to have been in academia for almost 40 years, meaning I've taught and conducted research at a large public university since 1969.

And I'm a "wannabe" consultant. From the start of my academic career I've had a passionate desire to translate research-based findings and principles into practical real-world procedures. For the past 30 years, I've fulfilled this desire by teaching useful principles and procedures to real-world practitioners and agents of change.

Over the years I've become a better consultant. I've continued to gain more research-based information, and I've learned through feedback how to communicate academic knowledge more effectively — from my oral presentations to written expression in books, magazines, and conference proceedings. But I want to improve further — at teaching practical research-derived knowledge to enrich the health, safety, and human welfare of clients I serve and the public in general. I am a "wannabe better" consultant.

Valuing the differences

I learned the hard way to adjust my academic stance when in the public forum. First, I found I needed to refrain from using complex rhetoric, or verbal behavior used to appear intellectual or professional. This is one of academia's biggest problems — the failure to disseminate research findings with language everyone can readily understand.

Simplistic language facilitates understanding and acceptance.

What a disappointment to have important principles and applications couched in a scholarly lingo that isn't readily digested by the public. That leaves the diffusion of information to the less-than-academic "pop psychologists" who often water down good information. But if these individuals don't put it out there, too often it would not get beyond the ivory towers of academia.

Next, I realized I should hold back on naming the "pop psychology" authors I was criticizing: It took the focus away from the point I was making. And I activated in the minds of my audience the names of the very persons I wanted them to forget.

In the academic world we reference everything that is not entirely original. We are careful to recognize the original source of information, even when challenging or discrediting certain aspects of that information. In contrast, consultants rarely reveal the source of material they use or discount, nor do they criticize other consultants or their material. They stick to presenting their own perspective without regard to the origins of that perspective.

I appreciate the consultants' avoidance of public critique of another consultant's procedures. Still, the academic in me wishes consultants would give credit to those individuals (usually academics) who first developed a particular approach. I've found audiences appreciate hearing where consultants learned the information they share.

One's competence as a consultant is not diminished by giving credit to the original source of a particular principle or procedure. Indeed, by referring to solid research supporting a particular intervention process, you very likely increase your own credibility. But try to find the primary research-based source rather than crediting an author who merely describes the work of someone else. If you have difficulty finding a primary source, contact an academic. Knowing the sources of information in the discipline is the academic's forte.

Additional differences

The table below summarizes these and additional distinctions between academics and consultants. These differences are not given to pit one against the other, nor to give one more status. I offer these distinctions to increase understanding, and perhaps stimulate beneficial collaboration and mutual learning between academics and consultants. Indeed, I'm convinced each profession can benefit from the strengths of the other.

Academics	**Consultants**
Ivory-tower language	Real-world language
Cite research to support points	Rarely cite research
Name those who don't agree	No name-calling
Open to many views	Focus on the view they sell
Support numbers with statistical tests	Report number without statistical tests
Conservative	Exaggerate to sell
Stick close to the data	Drift from the data
Learning-focused presentations	Motivation-focused presentations
Reference primary sources	Reference secondary sources

Distinctions between academics and consultants

As an academic, I must emphasize the differences listed here are biased. They are merely my own observations, obtained over a 30-year career of attempting to perform effectively in both worlds. I feel privileged to have played the role of both academic and consultant. I have learned from each profession.

The quality of my academic performance has improved as a function of my consulting experiences, and vice versa. For this reason, I believe each profession can benefit significantly from the other. Each can build from the strengths of the other, with each profession improving from the process.

How consultants could help academics

• Consultants can teach academics how to better present their theories and research findings to the public.

• Consultants can solicit real-world problems in need of research attention.

• Consultants can pose application-focused questions in need of empirical answers.

• Consultants can help academics find practical applications for their research findings.

• Consultants can inspire academics to be more application-oriented in their theories, research, and scholarship.

How academics could help consultants

• Academics can provide research-based rationale for selecting one intervention approach over another.

• Academics can increase a consultant's credibility by providing the primary source of a principle or procedure.

• Academics can teach consultants how to develop, administer, analyze, and interpret basic assessment devices like perception and culture surveys.

• Academics can show consultants how to interpret and use data from a statistical analysis.

• Academics can advise consultants what scholarship to read in order to improve particular aspects of their practice.

I sincerely hope readers will not view this presentation as an attempt to demean academia or the practice of consulting. I want readers to see this as a balanced discussion of strengths and limitations of each profession. More importantly, I hope you see ways academics and consultants can learn from each other, thereby becoming more effective at serving the health, safety, and welfare of those who pay their salaries.

Of course, this type of improvement is more about experiencing greater intrinsic rewards for making bigger and better differences in the human dynamics of work, play, and everything in between.

Understanding Personality

Introduction

Everyone loves to talk about personality. Society celebrates personality. Practical leaders, meanwhile, take their understanding of personality and use it to enrich their own performance, their relationships, and their culture.

Chapter Six examines the complexity of personality from a variety of perspectives, using a variety of tools. Dr. Geller begins in Part 1 by teaching an essential concept about personality, the difference between a personality trait (more or less permanent from birth) and state (fluid, depending on the situation).

Leaders beware: Attitudes, actions and decisions of your employees, supervisors and top managers can change — depending on whether you encounter them at the company picnic, in the parking lot, on the golf course or in a boardroom. But personality flexibility has its limits, and leaders cannot change genetics. Know when you are dealing with a trait and with a state of mind.

The common "Big Five" traits — openness, conscientiousness, extroversion, agreeableness, and neuroticism — are thought by many researchers to be immutable, for instance. "But leaders should realize there are ways to influence the extent a particular personality trait is manifested in behavior," says Dr. Geller in Part 2, which studies the flexibility of personalities.

Three distinct personality types, and their influence on safety, are covered in Parts 3, 4, and 5. Loyal fans of the "American Idol" TV show certainly know the overly-confident personality. They have too much of a good thing, oozing self-confidence, which in the workplace can be hazardous to their well-being. Safety pros are familiar with the "Superman Complex" and the "It won't happen to me" mindset. Dr. Geller shows in Part 3 how leaders can administer reality tests to set the stage for giving the supremely cocky worker some acceptable and applicable behavior-based feedback.

Everyone knows the Type A personality, the consummate corporate competitor. But do you know the difference between Type A behavior patterns and Type A emotions? In Part 4, Dr. Geller gives leaders a tool to assess both their own, and their team's, propensity for Type A behavior versus Type A emotion. High-scoring Type A behavior indeed creates safety risks by making people less mindful of the moment. High Type A emotion scores put people at a different risk: their emotions of anger and hostility, exemplified by "road rage," can lead to heart disease and possible death caused by heart disease.

Along with Type As, we've all come across "Nervous Nellies." As Dr. Geller describes in Part 5, high-anxiety people are energetic, high-strung, self-conscious, more nervous than average, and generally work under tension. They worry over money and business matters, and have difficulty keeping their mind on one thing.

Not exactly a recipe for safety on the job. But Dr. Geller shows how the anxiety trait can be good for safety. People who have an ongoing internal concern about the safety of themselves and others are more likely to do whatever it takes to prevent the occurrence of personal injury.

What color is your personality? Part 6 reviews the concepts of True Colors®, which can be used to assess personalities, show their diversity, and also how personality types overlap. Leaders regularly make these kinds of evaluations when forming teams and handing out assignments.

Another challenge leaders face, though they may not always realize it, comes when it is time to hand out rewards or penalties. Different strokes for different folks, advises Dr. Geller. It's doubtful many safety pros know the difference between BIS and BAS. Part 7 shows the value of understanding the behavioral inhibition system (BIS) and the behavioral activation system (BAS). Of course some personalities overlap the two systems, but the bottom line lesson for leaders: Some people are motivated by extrinsic rewards or penalties and need them; others are self-motivated and could be insulted by their use. "This reinforces the idea that you, as leaders, must know your people," says Dr. Geller.

Dave Johnson, Editor
Industrial Safety & Hygiene News

1 Understanding the "Big Five"

Clearly, leaders have much to gain from a better understanding of how personality characteristics affect safety performance. Indeed, you cannot effectively lead a People-Based Safety™ initiative without grasping the various facets of personality and their effects on safety. In the People-Based Safety™ model of ACTS — Acting, Coaching, Thinking, and Seeing — aspects of personality come under "Thinking" and "Seeing."[1]

Increasing risks

I believe certain personality characteristics increase the probability an individual will perform at-risk behaviors and experience an unintended injury. Other personality dimensions influence one's willingness to engage in injury-preventive behaviors. In other words, there are personality factors that contribute to injury proneness, and other factors that facilitate injury prevention.

Of course personalities are complex, and sometimes a personality factor can work both ways. One research study showed that people who scored high on a measure of personal control were more likely to take risks, making them more susceptible to injury. But these individuals were also more likely to follow safety precautions, making them less likely to suffer injury.[2]

States and traits

We can't get too far into a discussion of personality factors and safety performance without distinguishing between personality states and traits.[3]

I'm biased to the state perspective. This is the idea that personality is dramatically shaped by environmental context. That is, personality characteristics are not etched in stone, but fluctuate according to the situation. A person might be an optimistic success-seeker in recreational activities but be a pessimistic failure-avoider when it comes to work.

This idea that personality characteristics are fluid states was the focus of psychology when I attended graduate school in the late 1960s. Nurture was presumed more important than nature in determining human behavior. As such, circumstances and contingencies (such as safety policies) can be changed to overcome dispositional or personality deficits.

This perspective provided significant impetus to psychology as a discipline that can benefit the human condition. For if environmental factors have more influence on human behavior than biological factors, the environmental variables that are manipulated and studied by psychologists deserve preferential treatment.

This is what you find in most self-help books and audiotapes. Change conditions of your situation, follow certain self-management steps, and —

voilà! — you can improve your attitude, behavior and even career success. You can be whatever you want to be, proclaim some of these pop psychologists. It's a matter of changing your "state" of mind.

Limits of change

Of course we cannot become any kind of person we'd like to be, even if all the relevant environmental variables are lined up on our side. We do face physical limitations. Not everyone can dunk a basketball, or run a four-minute mile. And certain personality characteristics are indeed genetically determined and inherited, according to psychological research conducted in the past two decades.

A systematic comparison of identical twins reared apart after only five months of age, with twins raised together, showed that those twin pairs raised in the same home were not more similar than those raised separately in respect to various personality traits.[4]

This and similar research has led to the conclusion that genetics account for about 50 percent of individual differences in personality.

So the pendulum has swung back toward the nature side of the nature/nurture question in recent psychological research. This doesn't diminish environmental factors as being inconsequential. Behavior is determined by an interaction of the context in which we perform with personality characteristics. Sometimes situations are the primary determinant of behavior; sometimes personality shapes behavior.

Situations influence your person state.

The 600 students in my introductory psychology class are likely to sit quietly, listen, and take notes, regardless of their personalities. The large-class environment inhibits their individualism. But in a less restrictive and perhaps less intimidating social setting, these same students will "come out of their shells" and their behavior will likely vary greatly as their personality traits are allowed more freedom of expression.

The lesson for leaders? Be aware of how the attitudes, actions, and decisions of your employees, supervisors, and top managers might change — depending on whether you encounter them at the company picnic, in the parking lot, or in a board-room meeting.

Primary factors

Now to take our discussion of personality further, let's examine attempts to put personality types or dimensions into boxes or categories. Leaders in particular need to resist this temptation, this stereotyping.

Many of you are familiar with the Myers-Briggs classification of people along four bipolar personality dimensions: extroversion vs. introversion, sensing vs. intuition, thinking vs. feeling, and judging vs. perceiving.[5] This is an outdated approach that is not even mentioned in most contemporary psychology texts. Instead, every research-based textbook covering personality traits, among the more than 20 I consulted, identified the same five primary person factors.[6]

As a leader, you're sure to encounter these five primary types of personalities. The table below identifies these traits as bipolar dimensions, referred to in the research literature as the "Big Five." Note that the order of the dimensions presented in Table 1 spell the word "OCEAN," making it easy to remember these critical personality characteristics.

Openness to Experience
curious, broad interests, creative, original, imaginative, nontraditional, flexible, sensitive, adventuresome

conventional, down-to-earth, narrow interests, rigid, inflexible, insensitive, crude

Conscientiousness
achievement-oriented, organized, reliable, hard-working, careful, self-disciplined, ambitious, persevering, responsible

aimless, unreliable, lazy, careless, lax, negligent, weak-willed, hedonistic, impulsive, disorganized

Extroversion
sociable, assertive, talkative, optimistic, people-oriented, outgoing, fun-loving, affectionate

reserved, sober, cautious, quiet, aloof, task-oriented, shy

Agreeableness
soft-hearted, trusting, good-natured, helpful, forgiving, caring, cooperative, gentle

cynical, rude, suspicious, irritable, manipulative, vengeful, uncooperative, ruthless, hostile, self-centered, headstrong

Neuroticism
worrisome, nervous, emotional, insecure, hypochondriachal, frequent distress, hypersensitive, excitable

calm, relaxed, unemotional, hardy, secure, self-satisfied, composed

Bipolar dimensions of the "Big Five" personality traits

These dimensions remain relatively stable over an individual's lifetime, and carry across (generalize) cultural lines, according to substantial research conducted in the 1990s.[7] Plus, the Big Five personality traits have been found to be 40 percent to 60 percent inheritable.[8]

The table below provides a few representative questions per each of the Big Five traits. Higher numbers reflect qualities of the particular trait, except for those items followed by an "R". These need to be reverse scored, meaning the number circled should be subtracted from "6" to make the score consistent with other items.

Please don't get carried away trying to label people, putting them in one of the Big Five boxes through a scoring system. Don't consider your totals on these select items to be a valid measure of the Big Five. Rather, read them to improve your understanding of each trait, and use individual scores to facilitate conversations about these constructs and their relation to safety-related behaviors.

How do the "Big Five" affect safety performance?

We don't really know the answer to that — yet. Research to date has not studied specific connections between Big Five traits and injury proneness or injury prevention.

Still, certain relations between these personality traits and safety-related behaviors seem reasonable and worth the serious consideration of leaders:

For example, a case can be made for a positive correlation between anxiety and proactive injury prevention (I write in more detail on this connection later in this chapter), and the definition of anxiety I used is similar to the neuroticism dimension.

Plus, it is likely individuals scoring high on conscientiousness are more apt to partake in injury-prevention processes.

I also predict those scoring relatively high on openness to experience will

---

Openness to Experience
I have a vivid imagination.
I have a rich vocabulary.
I am not interested in abstract ideas. (R)

Conscientiousness
I am exacting in my work.
I neglect my duties. (R)
I like order.
I pay attention to details.
I am always prepared.

Extroversion
I feel comfortable around people.
I start conversations.
I keep in the background. (R)
I don't like to draw attention to myself. (R)
I am quiet around strangers. (R)

Agreeableness
I feel little concern for others. (R)
I make other people feel at ease.
I feel others' emotions.
I insult people. (R)
I take time out for others.

Neuroticism
I change my mood a lot.
I get upset easily.
I worry about things.
I get stressed-out easily.
I am relaxed most of the time. (R)

Sample questions to assess the "Big Five"

be more likely to accept and participate in an innovative approach to preventing injuries.

And common sense suggests injury-prevention procedures requiring interpersonal conversation (such as behavioral observation and feedback) will be more accepted by those who score high on extroversion and agreeableness.

It's also likely the "agreeable extroverts" will be more successful at implementing an interpersonal coaching process, a crucial component of leading People-Based Safety™.

These hypotheses reflect only a few of the possible ways the Big Five could influence workplace safety. Clearly, we need systematic research in this area, especially given the pervasive behavioral influence of these five genetically determined dimensions of human personality. As I mentioned at the outset, we have much to learn about personality that could benefit our safety initiatives in the workplace.

2 Can Personalities Change — For Safety's Sake?

A "safety and personality" presentation I once delivered at a professional development conference of the American Society of Safety Engineers was well attended and triggered a number of questions and comments, such as:
  • Aren't personality traits inherited and unchangeable?
  • How does personality affect perception and attitude — and safety-relevant behaviors such as "actively caring"?

I'd like to address these provocative questions. The answers will surely prove valuable to safety leaders and enhance their understanding of their workforce.

Aren't personality traits inherited and unchangeable?

In theory, a personality trait is permanent, especially when our surrounding environment (at home or at work, for example) does not prescribe a certain behavioral protocol. When we are free to express ourselves, our personality has a powerful influence on what we do.

But a critical question remains: Can personality and its impact on behavior be changed?

Most personality researchers and scholars claim certain personality characteristics (termed "traits") are essentially immutable, and cannot be targeted for change tactics. These include the very common Big Five traits which were introduced in the previous essay — Openness, Conscientiousness, Extroversion, Agreeableness, and Neuroticism.

Many personality researchers suggest we must accept the reality that people are born to express certain personality characteristics. But leaders should realize that there are ways to influence the extent a particular personality trait is manifested in behavior.

Someone who is naturally low on a Big Five trait can be influenced to express this characteristic through an environmental protocol (such as a policy), a behavior-change intervention (such as an incentive offering), or interpersonal dialogue (such as coaching).

Students in my large University classes, as well as participants in my professional development workshops, often inspire me to transition from my natural tendency to be shy and introverted to behave in an outgoing and extroverted manner.

To understand the potential flexibility of personality traits, I find it useful to consider how we use our hands. While most of us have a clear preference to use one hand over the other for specific activities, we can use the other hand when situations call for this change. It feels awkward, but we can do it. And with practice we can get quite good with our "off" hand.

Likewise, practice can make it feel natural to behave contrary to a personality trait.

How does personality affect perception and attitude?

Our personality influences our readiness to perform in certain ways. It makes us naturally aware or unaware of certain aspects of our life space. It influences how we interpret the various happenings in our daily lives. And personality affects how we respond to environmental stimuli, biasing our perceptions. We selectively attend to some things and screen out others. And as mentioned, environmental and social circumstances interact with our personality traits to enhance, neutralize, or inhibit them.

How can personality influence specific safety-related behaviors?

Narrowing our focus to only the Big Five, it's intuitive that people who score higher on extroversion and agreeableness are more people- and relationship-oriented by nature. They will also be more comfortable with safety procedures that involve interpersonal interaction and influence, such as coaching. Also, those displaying a high degree of openness will be more likely to approach new safety initiatives with an open mind, and will be less likely to resist change.

What personality types promote self-motivation?

This is certainly a major objective in safety — to move *from* "other-directed" behaviors that occur because we are held accountable by observers, supervisors, or coworkers *to* "self-directed" behaviors that occur because we hold ourselves accountable.

In an ideal, safety-mature culture, employees don't need outside accountability systems to motivate them to follow safety procedures. They hold themselves accountable to stick to the safety protocol when working alone, even in their backyards at home, when only the squirrels know if they're wearing eye protection while chopping wood or mowing the lawn.

From the Big Five, it's obvious conscientiousness is most aligned with self-accountability. But I also expect neuroticism to be related. Some degree of ongoing anxiety contributes to the self-motivation needed to keep a person doing the right thing for safety when working without supervision. I'm not talking about extreme neuroticism, but a level somewhere between "completely calm, relaxed, and unemotional" about an injury possibility and "nervous, emotional, insecure, and distressed" about safety issues.

I am only presenting you with intuitive hypotheses about a sample of personality characteristics identified through psychological research. Actual relationships between personality traits and workplace safety have not yet been systematically studied. As a leader you need to remember this and not jump to conclusions about what makes people tick.

This much is clear to me: leaders are well-served to increase their awareness and understanding of the role personality can play in injury proneness and injury control. And relationships between personality predispositions and voluntary participation in safety efforts are worthy of empirical study.

Possible Personality Connections to Safety

- **Conscientious** personalities are most aligned with self-accountability.
- **Neuroticism** — some degree of ongoing anxiety — contributes to the self-motivation needed to keep a person doing the right thing for safety when working without supervision.
- Personalities exhibiting **Extroversion** and **Agreeableness** are more people- and relationship-oriented by nature. They will be more comfortable with safety procedures that involve interpersonal interaction and influence, such as coaching.
- Personalities displaying **Openness** will be more likely to approach new safety initiatives with an open mind, and will be less likely to resist change.

3 Too Much of a Good Thing

"Believe and achieve" bellows a candidate on the popular TV game show "Deal or No Deal."

"We believe" scream the zealous fans at the college basketball game.

"Self-confidence is key to personal success" asserts the instructor of a leadership seminar.

"Self-affirmations enable you to reach your dreams" declares the keynote speaker at a professional development conference.

I bet most leaders have heard these or similar statements. In the academic world these motivational slogans reflect "self-efficacy" — the belief one can accomplish a certain task well. Research suggests clinical therapy can only be effective if the client has self-efficacy regarding the therapeutic process.

In other words, treatment cannot work unless the client believes it will work. The title of Albert Bandura's renowned 600-page text published in 1997 says it all: *Self Efficacy: The Exercise of Control.*[9]

Does all this sound like good common sense? Do you believe?

I want to play "devil's advocate" to this self-efficacy proposition. I'm not doing this merely for sake of argument, but to seriously challenge this staple of motivational speakers and clinical psychologists.

Here's my view: Too much self-efficacy or self-confidence can be self-defeating in some situations. When those situations involve risks or hazards, the result can be injury or death.

Self-confident "idol wannabes"

The popular reality TV show "American Idol" has millions watching one "idol wannabe" after another show off vocal talent (or lack thereof). Contestants travel long distances and wait in long lines for their chance to "strut their stuff." And it's clear from pre-performance interviews these contestants believe they have the "stuff." Many believe they could be the next "American Idol." Their self-confidence is at peak levels, perhaps "over the top."

Those of you who have seen performances of early "American Idol" contestants

Overconfidence can be undesirable.

realize where I'm going with this. Numerous candidates strut confidently on the stage, and then display no talent. Their performance is often humorous if not downright humiliating, and the judges do not refrain from laughing. How can people embarrass themselves like this on national television? Why didn't someone tell them they had no talent? Perhaps some contestants were misled by family and friends who gave them positive feedback to build their confidence and avoid hurt feelings.

Perhaps some candidates do not feel competent, but are only looking for three minutes of national attention. But many performers are visibly surprised and devastated by the judges' negative reactions. These individuals' extreme self-confidence put them in position to be publicly ridiculed and to become emotionally distraught.

Relevance to safety

The conceptual leap to industrial safety is neither difficult nor risky. It should be obvious to leaders that over-confidence on the job can put workers at-risk for injury. Leaders know extensive experience at a hazardous task (like driving in heavy traffic) can lead to an unhealthy degree of self-confidence or self-efficacy.

Many drivers, for example, gain so much confidence they add distracting behaviors to their driving routine, such as using a cell phone or fumbling for a CD. How about the common belief, "It won't happen to me"? Does this emanate from excessive self-confidence? Can the perception we are overly competent at a task put us at risk?

Reality check

Here's another common slogan: "Perception is reality." Perception may be reality for the individual, but there is a more accurate and valid reality out there. The self-confidence of the incompetent and foundering contestants on "American Idol" gave them a biased reality before their performance, but afterwards feedback gave them another reality. Likewise, risk-taking and distracted workers re-evaluate their realities after a near hit or injury. Some call this a "wake-up call;" I call it a "reality check."

Self-confidence and outcome feedback

The reality check after an embarrassing performance or a personal injury is too late. Yes, this outcome feedback will likely alter an individual's reality and inspire a need to change. But what kind of change is called for? And will the person accept the recommended change?

Leaders beware: Sometimes people are invested so much in a particular approach, or paradigm, they resist change. Self-confidence can fuel such resistance. In other words, a person's perception of self-effectiveness can inhibit facing the reality of a need to change. Another saying comes to mind:

"You can't teach an old dog new tricks."

After receiving devastating feedback from three judges, some contestants on "American Idol" declare their intent to try again next year. Their Teflon® self-confidence deflects the blow of unfavorable and unfriendly feedback. That's good news and a benefit of self-confidence. But the bad news is more costly and undermining. Excessive self-confidence can cause denial of the reality of failure. Instead it motivates a "stay-the-course" attitude of denial and resistance rather than a realistic re-evaluation of talents and resources, and openness to future possibilities.

Self-confidence and process feedback

Whether considering contest contenders or workers, I'm sure leaders see the special value of process feedback. People benefit more from feedback that pinpoints behaviors to continue and behaviors to eliminate than from simple outcome feedback that only evaluates the end result. Learning how well one accomplished a task is certainly useful, and can motivate or de-motivate subsequent performance. But knowing the end result (like where a golf ball lands) is not as helpful as knowing what behavior(s) can be improved, such as posture and follow-through.

Feedback at the start of learning is most influential.

The earlier people get feedback about the desirable and undesirable qualities of their behavior, the greater the acceptance and application.

This is common sense, right? When we are first learning a task we ask for process feedback, and if a correction is advised, we adjust accordingly. But after substantial experience at a task, process feedback often has less impact.

Leaders should understand the strong role of self-efficacy. Before we gain confidence and a sense of effectiveness at a task, we willingly accept and apply process feedback from a credible leader. After we become experienced and self-confident, process feedback can feel insulting. Now the credibility of the leader might be questioned. "Who are you to tell me how to improve? I've been doing it this way longer than you've been a safety coach."

To conclude

Leaders need to understand how self-confidence can go awry and inhibit continuous improvement. Such understanding can lead to the kind of reality testing that enables realistic aspirations, and sets the stage for acceptable and applicable behavior-based feedback.

Dealing with Overconfident Attitudes

- Give people feedback that pinpoints behaviors to continue and behaviors to eliminate.
- Don't rely on simple outcome feedback — safety scoring systems or injury rates — that only show the end result.
- Give people feedback about the desirable and undesirable qualities of their behavior as early as possible. Early on, before experience creates hard-to-change self-confidence, people are more open to accepting and applying your feedback.
- Your objective as a leader: understand how self-confidence can go awry and inhibit continuous improvement.

4 You Type A's: Slow Down for Safety

One personality factor with the most relevance to personal injury and its prevention is Type A versus Type B. Type A individuals are more prone to experience a near hit or unintentional injury. On a personal note, I realize my Type A propensities make it difficult for me to live in the moment and attend mindfully to my ongoing behavior and its environmental context.

What is Type A?

Most leaders have heard about the Type A personality; in fact, many leaders are Type As. The Type A personality type was identified in the late 1950s by physicians Meyer Friedman and Ray Rosenman as a pattern of behavior presumed to contribute to heart disease.[10] Type As are competitive, impatient, hostile, and always striving to do more in less time. In contrast, Type Bs are calmer, more patient, less hurried, and less hostile.

Early studies of the Type A personality evidenced a positive correlation with heart disease. Specifically, Type A people were more likely than Type B people to have heart attacks. This association became well known in the 1960s and 1970s, and is still referenced in contemporary self-health books. But more recent research casts doubt on this conclusion, as I will show later.

The positive correlation between Type A behavior and unintentional injury remains apparent, whether referring to empirical research or common sense. People who are impatient and continually attempt to do more in less time are more likely to hurt themselves or others. With a future-oriented mindset, Type A individuals short-circuit the moment in favor of planning ahead. They're mindful of their next moves, but they rush past the pleasures of the present.

Type A persons are often impatient.

Who are you?

OK, I've owned up to my Type A behavior patterns, and realize my special challenge to slow down and become more mindful and appreciative of the present. What about you?

Do you get impatient and experience negative emotions when driving behind a vehicle traveling the speed limit in the left-hand lane?

Do you look for the shortest line in the grocery store by estimating the number of items in others' carts? And if another line is moving faster, do you quickly switch to that line? Then, do you feel angry when you notice the line you just left starts to move faster than your new line?

Do you walk on the moving sidewalk in airports, even though you have plenty of time before your next flight? And do you feel any hostility toward those laid-back Type B people blocking your path, as you hustle to a departure gate or the baggage-claim area — only to wait impatiently for another flight, checked luggage, or a cab?

Type A behavior vs. Type A emotion

Notice that these examples of Type A people reflect both behavior and emotion. In other words, while you're rushing to

Type A behavior affects Type A emotions, and vice versa.

save time (behavior), how do you feel about people who get in your way and slow you down (emotion)?

While the hurried behavior increases risk for personal injury, certain distressing emotions put people at risk for heart disease. Specifically, the emotions of hostility and anger that often accompany time-shaving behavior relate to the development of heart disease.[11] It's critical to distinguish between Type A behavior and Type A emotion. Type A behavior puts people at risk for unintentional injury, but not for heart disease. Type A emotions of anger and hostility, exemplified by "road rage," put people at risk for heart disease and for death following heart disease.

Survey to Assess Type A Behavior and Type A Emotion

Type A Behavior

___ 1. I experience a lot of time pressure.

___ 2. I feel the pressure to get ahead and succeed.

___ 3. I do many things fast — talking, walking, eating, and so forth.

___ 4. I work long hours.

___ 5. I often want to (and sometimes actually do) finish other people's sentences.

___ 6. I think it's important to acquire a lot of money and possessions.

___ 7. I hate to stand in line.

___ 8. It is not necessary for others to impose deadlines; I set them for myself.

___ Total

Type A Emotion

___ 1. People seem to annoy me intentionally.

___ 2. I often raise my voice.

___ 3. Many situations make me angry.

___ 4. People consider me short-tempered.

___ 5. I have a "short fuse" when it comes to tolerating incompetence.

___ 6. Being caught in slow traffic frustrates me so much I want to yell at the other drivers.

___ 7. I try to control my temper, but often lose it.

___ 8. I express my anger physically by hitting, kicking, slapping, or throwing things.

___ Total

Assess yourself

The sidebar above includes items you can use to measure your propensity for Type A behavior versus Type A emotion. Simply write a number from one to seven next to each item to estimate the extent to which the statement applies to you. A "1" reflects "not at all," a "4" indicates "sometimes," and a "7" designates "all of the time." The higher your score for the Type A behavior items, the more difficult it is for you to be mindful of momentary risks to your safety. A relatively high score for the Type A emotion items suggests risk of heart disease.

A score above 30 on either scale suggests Type A tendencies. I score 40 for Type A behavior and 22 for Type A emotions. My high Type A behavior tally supports my earlier confession of being injury-prone.

What about you?

As a leader, it would be worthwhile for you to give this scale to a work team and openly discuss personal scores. It could be quite enlightening to discover your relative ranking of Type A behavior versus Type A emotion. But most beneficial is a discussion of circumstances and contexts that influence Type A behaviors and emotions. This can lead to environmental changes that make it easier for more people to slow down and live in the moment. And to have an organization full of people willing to slow down and be mindful of the moment is certainly an important way to enrich your culture.[12]

5 Anxious About Hazards?

As a safety leader, you can place yourself — and your managers, supervisors and employees — in one of four person-state categories by considering whether your approach to safety is seeking-success or avoiding-failure.

Exclusive *success-seekers* are the most optimistic, while exclusive *failure-avoiders* are most pessimistic.

Overstrivers want success and abhor failure; while extremely motivated, they often feel distressed and insecure.

Although the *failure-accepters* are most unmotivated and apathetic, they are typically more content than the failure-avoiders and overstrivers.

States vs. traits

Leaders must be cautious when using this four-way classification system.

First, the motivational state of an individual varies from one situation to another and from one time to the next. A certain manager could be a failure-avoider for safety, a success-seeker for productivity, and an overstriver for quality. And one serious injury could move a person from failure-avoider to failure-accepter for safety.

I want to introduce a personality trait that can influence which of the four success/failure states an individual is experiencing, and whether a transition to a more healthy or productive state might be expected. I'm referring to a trait rather than a state — an enduring personality characteristic a person brings to every situation and influences one's state and relevant behavior.

The trait I am considering here is general anxiety, as measured 50 years ago by a highly researched survey instrument developed by Dr. Janet Taylor Spence. It's called the Taylor Manifest Anxiety Scale (TMAS).[13]

What is anxiety?

My *American Heritage Dictionary* defines anxiety as "a state of uneasiness and distress about future uncertainties." Clearly, anxiety is an unpleasant state we want to avoid. But here I'm not talking about a state, rather a trait. As a trait, anxiety is a relatively stable personality quality that determines one's motivational and emotional responsiveness to a particular circumstance. If the situation is anxiety-provoking, state anxiety combines with trait anxiety to produce a most unpleasant and distressing situation.

As measured by the TMAS, high-anxious people are energetic, high-strung, self-conscious, more nervous than average, and generally work under tension. They worry over money and business matters, and have difficulty keeping their mind on one thing. Regarding the success/ failure typology defined above,[14] high-anxiety individuals worry about being unsuccessful. They tend to be overstrivers or failure-avoiders.[15]

Prior experience can make certain situations anxiety-provoking.

At first, high anxiety seems to be an undesirable personality trait. But this is not necessarily so. Before bemoaning your own proneness toward high anxiety, consider the following reliable research finding.

While high-anxious people perform less competently than low-anxious people on novel tasks or jobs they could not prepare for, they typically outperform those with lower anxiety on tasks for which they could prepare.[16]

Why? Because their strong desire to avoid the aversive anxiety feelings accompanying failure motivates them to do as much as possible to succeed. With proper proactive preparation, these individuals become success-seekers.

An illustrative anecdote

Many of you have experienced test-taking situations you could not prepare for, like the Scholastic Achievement Test (SAT) high-school students take for entrance into college. On average, high-anxious individuals do worse than low-anxious people on these tests.

Why? Because their nervousness and heightened arousal energizes a wide range of behaviors, many of which compete with execution of the correct response.

Perhaps you won't be surprised to learn that SAT scores are often not predictive of academic performance in college, as measured by grade point average. Why? Because students can prepare for tests that determine their grades, and high-anxious students are typically very proactive when given opportunities to prepare for success. Their strong need to avoid the negative emotions of being unprepared motivates them to work hard to avoid failure. And if they prepare well and become confident they can make good grades, they can develop an achievement mindset and become success-seekers.

Relevance to safety leadership

I hope you agree the anxiety trait defined here can be good for safety. Recall the dictionary definition of anxiety as a state of uneasiness about the future. People who have constant anxiety about the possibility of a workplace injury are going to do everything they can to put themselves in control of preventing injuries, and so put their safety-focused anxiety on hold. These folks do not need an actual injury or fatality to get their attention. However, such unfortunate events are often a necessary wake-up call for the low-anxiety employees.

The term anxiety carries negative baggage in our culture, so the premise that anxiety is good for safety might be difficult for some leaders to accept. If so, try substituting the analogous term "concern" for "anxiety." The bottom line is that people who have an ongoing inherent concern about the safety of themselves and others are more likely to do whatever it takes to prevent the occurrence of personal injury.

Isn't this the kind of personality trait we hope to find among our safety leaders? These are the passionate safety leaders who are most likely to cultivate a state of anxiety or concern for safety throughout a work culture. Strange as it might sound, cultivating anxiety or concern, if you can properly moderate the level, is another way of enriching your culture and making it safer.

Know Your Workforce

- Success-seekers — optimistically work to achieve success
- Overstrivers — driven to avoid failure by working in overdrive to succeed; extremely motivated but often insecure
- Failure-avoiders — motivated by fear of failure, a sense of pessimism
- Failure-accepters — expects failure regardless of personal effort; most apathetic or indifferent

6 "Coloring" Your Team (and Yourself)

Several years ago, I had one-on-one conversations with top executives of a large Fortune 100 company. Each VIP office was spacious and uniquely luxurious. I did notice one common feature — on every mahogany desk sat a brass plaque with the executive's name and a colored circle. After three interviews, I realized each color on the three nameplate circles was different.

Sure, I was curious about these colored circles, and I asked my host to explain. He said each color identifies one of four distinct perspectives and personalities. In other words, each color represents the person's true character. He then went on to describe the personality characteristics of each color — *Orange, Green, Blue,* and *Gold.*

At my host's company, people are assigned to project teams according to their primary personality color. Why? He and his colleagues are convinced the most productive and synergistic teams include representation from each of the four personality colors. I review these four person traits later, but first let's review the basis of this four-color concept and how the four personality types are measured.

Personality characteristics

If you were asked to describe your personality, you could undoubtedly list a number of unique qualities you perceive in yourself as compared to others. You might describe particular physical characteristics, likes vs. dislikes, interests, motives, life goals, attitudes, feelings about life events, skills, abilities, and even typical behaviors. The list could be endless.

And you could also list a number of behaviors, attitudes, and life events that are influenced by your personality, and vice versa. In other words, we cannot deny certain aspects of our personalities — attributes, dispositions, and tendencies — influence our ongoing actions and our interpretations of those actions. Personality theory includes defining a generic basis for describing one's personality and then developing a reliable and valid approach to assessing meaningful variations among individuals. A meaningful personality difference is one that predicts reliable differences in people's behavior. Needless to say, this can be a valuable tool for leaders, first to assess their own personality, and then the diverse nature of their fellow workers.

Four personality colors

The most prominent personality theorists and researchers, from Hippocrates to Carl Jung and Myers/Briggs, have classified people into four groupings.[17] Professor of psychology David Keirsey, for example, developed the Keirsey Temperament Sorter, which identifies individuals as one of four different types of temperaments. In 1978, Don Lowry, a student of Keirsey's, introduced the notion of four personality colors. Lowry set about developing a fundamental way to package the information into guidelines that could be easily applied by children and adults. Also in 1978, Lowry founded True Colors, Inc., a program that identifies personality and career types according to the archetypes set forth by the Keirsey Temperament Sorter. True Colors® makes Keirsey's psychological theory more "real-world" by assigning each personality type to a corresponding color.

The True Colors® program has workshop participants identify themselves using four colors: Blue, Gold, Green or Orange. Each color has particular strengths and each analyzes, conceptualizes, understands, interacts and learns differently.

In the spring of 2004, LBC Global (now True Colors International) acquired True Colors®. Thousands of teachers, human resource managers, parents, students, and corporations have attended seminars, stage shows, the True Colors® University; read related books; and become certified as True Colors® facilitators.[18]

By showing us how we are inherently different from others, Lowry's concept of personality colors facilitates mutual appreciation and support of people's values, attitudes, and behaviors. Enthusiasts and facilitators of this approach to categorizing personalities claim effective work teams need the full spectrum of primary colors in order to maximize productivity and attain synergy.

Grouping desirable attributes

A particularly attractive quality of this approach of "coloring" one's personality is the use of only positive characteristics. Unlike most other personality theories and assessment techniques, this strategy does not identify negative or unwanted qualities of people. The focus is on people's positive or esteemed distinctions. According to Wikipedia, here is a brief sketch of Lowry's four-color system (with particular attention to how the colors would play out on a safety team or group):

Golds are practical and sensible, and favor structure and organization. Regarding industrial safety, Golds are detail-oriented and appreciate the need for governmental rules and regulations (i.e., OSHA and MSHA), and they support the traditional "discipline" approach to mandatory compliance.

Blues are empathetic, reflective, aware people. They believe in loyalty and belongingness. I believe Blues are most receptive to People-Based Safety™ because of their attention to person factors beyond behavior, including emotions, feelings, interpersonal trust, belongingness, and actively caring.

Greens are determined and persistent. They are apt to resist change — if change is not supported by theory and data. They tend to value information above feelings in making decisions. In terms of safety, you'd want a mix of Greens on your team to evaluate ideas and suggestions on the basis of the evidence — not "flavor of the month" marketing hype.

Oranges are competitors. They have a strong urge to act now, to win, to succeed. In terms of your safety committee or team, you would want to include some Orange types because they tend to throw themselves into projects, and they would work hard for the project's success.

Personality trait vs. state

I've discussed the critical distinction between person trait vs. state in *People-Based Safety™: The Source*.[19] This differentiation is clearly relevant for this discussion of Lowry's four colors. Based on the same theory and research as the famous Myers/Briggs Type Indicator,[20] this color categorization is considered a trait approach to personality. In other words, it's assumed a person's relative ranking of the four colors is consistent across situations and throughout one's life.

While the four-color typology seems useful to explain and resolve differences among individuals, and to assess team composition and balanced leadership qualities, I believe it's chancy to assume one's primary color is constant across settings. It's likely many people alter their color rankings to fit current circumstances. But I must admit my own color ranking — Orange, Blue/Green (tied), and Gold — is quite impervious to situational and interpersonal change. Still, I do find myself adjusting natural tendencies in order to be appropriate or relevant for a certain occasion.

Daily application

Besides private amusement, what practical benefit can come from learning your own or someone else's ranking of their personality colors? First, I have found it useful to bring color language into interpersonal conversations. For me, it's been fun to predict (privately) the colors rankings of my students, friends, family members, and colleagues. I'm hitting about 90 percent accuracy at predicting an individual's primary color, and most have confirmed their personality matches at least their highest and lowest ranked colors.

Enlightening conversation

Whenever I do not accurately predict a person's primary color, I activate intriguing conversation and gain fascinating information. For example, I judged my chiropractor's primary color to be Orange, because he has a very active and adventuresome lifestyle, from routine biking and weight-lifting to frequent kayaking and rock climbing.

A primary "blue" can be active and adventuresome.

Our conversation revealed, though, his primary color of Blue was most consistent with both his profession and general outlook toward people and situations. While engaging in active recreation that pushes the safe-and-secure limits, my chiropractor believes he is not a risk taker because he takes every possible safeguard, and he never competes at his sport. Indeed, he takes great delight in introducing others to the individual sports at which he is exceedingly competent (i.e., a Green attribute). He does this not to compete or show off his skills, but to share the exhilarating experience with others.

Colorful evaluations

I periodically inject color language in conversations with my graduate students (who have all taken two assessments of their colors). I often find myself evaluating a certain decision or behavior by saying to myself or to others, "Oh, that's the Orange in me." Or, I might say to someone, "That's mighty Blue of you." Thus, this colors schema provides straightforward and meaningful terminology for identifying and appreciating some of the person-based dynamics of human behavior.

Estimating sources of frustration

When my graduate students mention their frustrations and concerns, I can often see relationships between their primary color and their verbal behavior. Golds, for example, are frustrated by lack of organization, tardiness to important meetings, unfairness, unexpected events, and incompetence. In contrast, Blues are distressed by lack of empathy and sensitivity to others, and by judgmental, aggressive, non-communicative, and non-caring behavior.

The Orange style is frustrated by boredom, predictability, lack of humor, whining, nagging, time constraints, "couch potatoes," and slow behavior. On the other hand, Greens are distressed by incompetence, impulsivity, off-task distractions, ill-informed and/or illogical decision-making, and blind acceptance of the status quo.

Finding the right recognition

In *People-Based Safety™: The Source,*[21] I outlined the steps for giving behavior-based recognition, which included giving a universal statement of appreciation after identifying the specific behavior you want to acknowledge. But what kind of general comment should you make?[22]

It can be useful to consider an individual's primary color. Here are some possible ways to acknowledge accomplishment per a person's primary color.

Golds should like to hear "I affirm your integrity and sincerity;" "Your sense of duty and personal responsibility is noticed and highly regarded;" and "Your efficiency, dependability, and loyalty to our organization are admirable."

In contrast, Blues would appreciate, "Your actively caring for others is greatly appreciated;" "Your compassion for your coworkers is invaluable, and your ability to see potential in others is impressive;" "Your interpersonal and group communication makes a difference."

If the person is Orange, consider the following universals: "You are a leader who puts ideas into action;" "I appreciate your ability to take charge of our group and make things happen;" "I admire your passion and enthusiasm to motivate us to go beyond the call of duty."

Contrast these Orange-directed kudos with special ways to recognize contributions from Greens: "Your analytical abilities are invaluable;" "Your benchmarking and critical evaluation enable us to shoot for world-class;" "Thank you for teaching us to substitute objective data for subjective opinion;" "Would you be willing to share your enlightening observations with the entire company?"

Despite the fun you can have "coloring" yourself and others, and its potential application to safety teamwork, I do not want you to label yourself as a particular color, or to pigeonhole yourself or others into a certain personality type. This is not about putting people into stereotypical boxes. Rather, I hope you use this discussion of Lowry's personality colors to appreciate the diversity of people's natural talents and propensities to act in certain ways. I hope you view such heterogeneity as enriching a work culture and facilitating the synergistic outputs expected from world-class organizations.

7 Different Strokes

The articles in this chapter on "Understanding Personality" have introduced specific personality factors that can impact industrial safety. Part 5 discusses the relationship between anxiety and injury prevention, with reference to a classic measure of high vs. low anxiety. Part 1 introduces the five personality factors discussed most consistently in current textbooks on basic psychology — Openness to experience, Conscientiousness, Extroversion, Agreeableness, and Neuroticism (OCEAN). Representative questions from an assessment device used to measure each of these person factors are included.

Part 4 addresses the well-known Type A personality trait, explains a distinction between Type A *behavior* and Type A *emotion*, and reveals how this personality factor influences injury-proneness.

Part 6 covers the True Colors® approach to understanding four basic personality types, from assessment to practical application.

Now this article targets yet another person factor relevant to the human dynamics of injury prevention — individual sensitivity to rewards vs. penalties.

Evidence-based theory

Working initially with animals, physiological psychologists have identified two distinct neurological structures and systems — the behavioral inhibition system (BIS) and the behavioral activation system (BAS).[23]

The BIS is sensitive to signals of possible negative consequences and inhibits behavior that may lead to undesirable outcomes.

In contrast, the BAS is sensitive to cues for rewarding consequences, and facilitates the initiations of behavior that can gain positive consequences.

Bottom line: The BIS is presumed to regulate avoidance behavior, while the BAS controls approach behavior.

Researchers connect the BIS to anxiety or fear of failure and the BAS to impulsivity and a need to achieve soon, certain, and positive consequences. I address the need to avoid failure versus achieve success in *People-Based Safety™: The Source*.[24] Here we're relating these distinct motivational perspectives to independent neurological systems. Plus, physiological psychologists have related specific brain activity in humans to differential BIS versus BAS sensitivities.

In other words, people vary significantly with regard to their susceptibility to positive reinforcement versus negative reinforcement and punishment — an important point for leaders to consider.

Distinguishing sensitivity

Researchers have developed and tested a psychological survey that measures an individual's *independent* BIS versus BAS motivational systems. The term "independent" is important here, because a person's sensitivity to BIS does not influence sensitivity to BAS. An individual can score high on both dimensions, reflecting significant impact of both positive and negative consequences. This outcome is analogous to the *overstriver* who is simultaneously motivated to both achieve success and avoid failure.

A *success-seeker* would score high on BAS and low on BIS, while a *failure-avoider* would score low on BAS and high on BIS. Of course, it's possible to score low on both the BAS and the BIS, possibly indicating low motivation to achieve positive consequences or avoid negative consequences.

Some people are sensitive to both winning and losing.

It's tempting to place this type of person in the *failure-accepter* category of the 2x2 matrix defined by answering yes or no to two questions: 1) Do you seek success? and 2) Do you avoid failure? But this classification could be incorrect and misleading.

The BIS and BAS scales measure people's attention to external or extrinsic consequences. However, some people perform more for intrinsic than extrinsic consequences. They are not controlled by external contingencies. These individuals are self-directed. They set personal goals and reward themselves internally when they achieve. They might also perceive significant intrinsic or natural reinforcers for their behavior So leaders beware: a low BIS and BAS score does not necessarily indicate lack of motivation, but might simply reflect predominance of self-directed over other-directed behavior.

An assessment of BIS and BAS

Have I piqued your interest in this particular personality dimension? Perhaps you're curious about your sensitivity to extrinsic positive and negative consequences. The 12 questions in the sidebar were selected from the research-based assessment tool of 36 BIS questions and 31 BAS items.[25] Although these are the items showing the greatest statistical connection to the measured construct — BIS or BAS — they do not provide a valid measure of your propensity to be influenced by positive and negative consequences. Still, you can answer the questions to estimate your BIS and BAS sensitivities and to increase your understanding of this personality construct.

BIS or BAS?

Please answer "yes" or "no" to each of the following questions:

1. Do you often refrain from doing something because of your fear of being embarrassed? ___ No ___ Yes

2. Do you like being the center of attention at a party or a social meeting? ___ No ___ Yes

3. Do you, on a regular basis, think you could do more things if it were not for your insecurity or fear? ___ No ___ Yes

4. Do you often do things to be praised? ___ No ___ Yes

5. Are you easily discouraged in difficult situations? ___ No ___ Yes

6. Do you sometimes do things for quick gains? ___ No ___ Yes

7. Are you often afraid of new or unexpected situations? ___ No ___ Yes

8. Do you like to put competitive ingredients in your activities? ___ No ___ Yes

9. Are you often worried by things you said or did? ___ No ___ Yes

10. Do you generally give preference to those activities that imply an immediate gain? ___ No ___ Yes

11. Whenever possible, do you avoid demonstrating your skills for fear of being embarrassed? ___ No ___ Yes

12. Does the possibility of social advancement move you to action, even if this involves not playing fair? ___ No ___ Yes

Scoring the BIS/BAS scale

Your BIS and BAS scores can be readily obtained. As you might have guessed, the six odd-numbered questions assess BIS, and the even-numbered items measure BAS. So, total the number of "yes" items separately for the odd and even-numbered questions. This gives you a BIS and a BAS score. The higher the score, the greater your sensitivity to control by extrinsic consequences — the BIS total for penalties and the BAS total for rewards.

Interpreting scores

What do your totals for the BIS and BAS items mean?

First, norms have not been established for the entire BIS/BAS survey, let alone the selected items given here. It's impossible to know whether your scores are high or low relative to a meaningful standard. Still, you can get some indication of your relative sensitivity to negative versus positive consequences.

More than three "yes" responses to the BIS and/or BAS items suggest substantial susceptibility to the type of extrinsic consequence implied. If the difference in your BIS vs. BAS totals is three or more, a differential sensitivity to one type of consequence is indicated. Of course, you might score high on both BIS and BAS items, indicating equivalent motivation to achieve success and avoid failure. This characterizes the overstriver, as discussed above.

A low number of "yes" responses for both the BIS and BAS does not necessarily indicate low motivation. As entertained earlier, this could imply a self-directed individual motivated by intrinsic reinforcers and internal behavior management.

Connections to occupational safety

How does the BIS/BAS sensitivity distinction relate to the human dynamics of occupational safety? I'm sure you see a number of connections.

First, it shows individual diversity along with a critical aspect of behavior management. Some people are more sensitive to punitive than positive consequences and vice versa. We can't expect everyone to be equally influenced by the same behavior-based contingencies. It's natural for some managers to choose one type of consequence over another to motivate behavior, and it's instinctive that some workers are more motivated by negative over positive consequences, and vice versa.

What about those who are surprisingly indifferent to both positive and negative consequences? It's possible these individuals simply do not care about safety and/or their job, and should be asked to leave. But it's also possible these folks are internally motivated and don't need external consequences to keep them going. These individuals can actually be insulted and de-motivated by an explicit attempt to control their behavior with extrinsic rewards, which they might perceive as bribes. This reinforces the idea that you, as leaders, must know your people.

To conclude

People are uniquely different in many ways, including their sensitivity to extrinsic behavior-based contingencies. This implies a need for leaders to find out more about person factors before implementing a policy or intervention process to improve safety-related behavior. The BIS/BAS survey items provided here could stimulate valuable group discussion before a behavior management intervention is designed. They also suggest questions for leaders to ask individuals who appear disinterested or uninfluenced by certain consequence-based interventions implemented to prevent occupational injuries.

Endnotes

Chapter 1

1. Skinner, B. F. (1953). *Science and human behavior.* New York: Macmillan.
2. Skinner, B. F. (1971). *Beyond freedom and dignity.* New York: Alfred A. Knopf.
3. Geller, E.S. (2005). *People-based safety™: The source.* Virginia Beach, VA: Coastal Training Technologies Corporation.
4. For a more complete explanation, read "Don't over-justify safety," p. 86, in *People-based safety™: The source.*
5. See my discussion of "How to celebrate safety success," p. 163 in *People-based safety™: The source.*
6. I discussed this distinction in "Never stop teaching," p. 42 in *People-based safety™: The Source.*
7. Pounds, J. (2005). *Praise for profit: How rewards and incentives are demotivating America's workforce.* www.praiseforprofit.com.
8. Gilbert, T. F. (1978). *Human competence – Engineering worthy performance.* New York: McGraw-Hill.
9. Bailey, J.S., & Burch, M.R. (2005). *How to think like a behavior analyst.* Mahwah, NJ: Lawrence Erlbaum Associates.
10. Pounds, J. (2005). *Praise for profit: How rewards and incentives are demotivating America's workforce.* www.praiseforprofit.com.
11. Allesandra, T., & O'Connor, M.J. (1996). *The platinum rule: Discover the four basic business personalities and how they can lead you to success.* New York: Warner Books Inc.
12. Rogers, C. (1977). *Carl Rogers on personal power: Inner strength and its revolutionary impact.* New York: Delacarte.
13. Geller, E.S. (2005). *People-based safety™: The source.* Virginia Beach, VA: Coastal Training Technologies Corporation.
14. I discuss ways of building self-accountability in "Minding your own business," p. 185 in *People-based safety™: The source.*
15. Bem, D. J. (1972). Self-perception theory. In L. Berkowitz (Ed.), *Advances in experimental social psychology* (Vol. 6) (pp. 1-60). New York: Academic Press.
16. Skinner, B. F. (1971). *Beyond freedom and dignity.* New York: Alfred A. Knopf.
17. See "Create internal tension," p. 201 in *People-based safety™: The source.*
18. Allesandra, T., & O'Connor, M.J. (1996). *The platinum rule: Discover the four basic business personalities and how they can lead you to success.* New York: Warner Books Inc.
19. See "How we perceive risk," p. 222 in *People-based safety™: The source.*
20. Bem, D. J. (1972). Self-perception theory. In L. Berkowitz (Ed.), *Advances in experimental social psychology* (Vol. 6) (pp. 1-60). New York: Academic Press.

Chapter 2

1. Collins, J. (2001). *Good to great.* New York: Harper Collins.
2. ibidem, p. 50.
3. See "Safety needs your leadership," p. 272 in *People-based safety™: The source.*
4. Collins, J. (2001). *Good to great.* New York: Harper Collins, p. 89.
5. Deming, W. E. (1986). *Out of the crisis.* Cambridge, MA: Center for Advanced Engineering Study, Massachusetts Institute of Technology.

6. Collins, J. (2001). *Good to great.* New York: Harper Collins, p. 142.

7. For more on the success-seeker vs. failure-avoider perspective see "Thinking and personality," p. 204 in *People-based safety™: The source.*

8. Collins, J. (2001). *Good to great.* New York: Harper Collins.

9. ibidem, p. 28.

10. Miller, D. T., & Ross, M. (1975). Self-serving biases in attribution of causality: Fact or fiction? *Psychological Bulletin, 82,* 313-325.

11. Collins, J. (2001). *Good to great.* New York: Harper Collins, p. 210.

12. Daniels, A. C., & Daniels, J. E. (2005). *Measure of a leader: An actionable formula for legendary leadership.* Atlanta, GA: Performance Management Publications.

Chapter 3

1. Covey, S.R. (2004). *The 8th habit: From effectiveness to greatness.* New York: Simon and Schuster, Inc.

2. ibidem

3. Covey, S.R. (1989). *The seven habits of highly effective people.* New York: Simon and Schuster.

4. Covey, S.R. (2004). *The 8th habit: From effectiveness to greatness.* New York: Simon and Schuster, Inc, p. 270.

5. *The American Heritage Dictionary* (1991). Boston, MA: Houghton Mifflin Company, p. 1240.

6. ibidem, p. 1146.

7. Covey, S.R. (2004). *The 8th habit: From effectiveness to greatness.* New York: Simon and Schuster, Inc, p. 84-85.

8. Farber, S. (2004). *The radical leap: A personal lesson in extreme leadership.* Chicago, IL: Kaplan Publishing; Farber, S. (2006). *The radical edge: Stoke your business, amp your life, and change the world.* Chicago, IL: Kaplan Publishing.

9. Covey, S.R. (2004). *The 8th habit: From effectiveness to greatness.* New York: Simon and Schuster, Inc.

10. Geller, E.S., & Lehman, P.K. (Eds.) (2007). *Teaching excellence at a research-centered university: Energy, empathy, and engagement in the classroom.* Boston, MA: Pearson Custom Publishing.

11. Behavior Safety Now is an annual international conference supporting the Cambridge Center for Behavioral Studies. The 2006 BSN conference was held Oct. 2–5 in Kansas City, KS.

12. ibidem

13. George, W. (2003). *Authentic leadership.* San Francisco, CA: Jossey-Bass.

14. Petersen, D. (2001). *Authentic Involvement.* Chicago, IL: The National Safety Council.

15. George, W. (2003). *Authentic leadership.* San Francisco, CA: Jossey-Bass, p. 12.

16. *The American Heritage Dictionary* (1991). Boston, MA: Houghton Mifflin Company, p. 300.

17. Petersen, D. (2001). *Authentic Involvement.* Chicago, IL: The National Safety Council, p. 46.

18. Deming, W.E. (1991, May) *Quality, productivity and competitive position.* Workshop presented by Quality Enhancement Seminars, Inc., Cincinnati, OH.

19. Krause, T.R. (2005). *Leading with safety.* Hoboken, NJ.: John Wiley & Sons, Inc.

20. Myers, I. B., & McCaulley, M. H. (1985). *Manual: A guide to the development and use of the Myers-Briggs Type Indicator.* Palo Alto, CA: Consulting Psychologists Press.

21. Daniels, A. C., & Daniels, J. E. (2005). *Measure of a leader: An actionable formula for legendary leadership.* Atlanta, GA: Performance Management Publications.
22. Geller, E. S. (1996). *The psychology of safety: How to improve behaviors and attitudes on the job.* Radnor, PA: Chilton Book Company.; Geller, E. S. (2001). *The psychology of safety handbook.* Boca Raton, FL: CRC Press.
23. See "Thinking and Personality", p. 204 in *People-Based Safety™: The source.*

Chapter 4

1. Blanchard, K., & Johnson, S. (1982). *The one-minute manager.* New York: Simon & Schuster, Inc.
2. Pounds, J. (2005). *Praise for profit: How rewards and incentives are demotivating America's workforce.* www.praiseforprofit.com.
3. ibidem, p. 52
4. ibidem, p. 59
5. Carnegie, D. (1936). *How to win friends and influence people* (1981 ed.) New York: Simon & Schuster.
6. Blanchard, K. (1999, November). *Building gung ho teams: How to turn people power into profits.* Workshop presented at the Hotel Roanoke, Roanoke, VA; Blanchard, K., & Bowles, S. (1998). *Gung ho! Turn on the people in any organization.* New York: William Morrow and Company, Inc.
7. Ehrenreich, B. (2001). *Nickel and dimed.* New York: Henry Holt and Company, LLC.
8. ibidem
9. ibidem
10. ibidem
11. Abernathy, W.B. (1996). *The sins of wages.* New York: PerfSys Press.
12. Renwick, G. (1991). *A fair go for all: Australian/American interactions.* Yarmouth, ME: Nicholas Brealey Publishing.
13. Cialdini, R. B. (2001). *Influence science and practice.* (4th Edition). Boston: Allyn and Bacon.
14. ibidem
15. Lehman, P.K., & Geller, E.S. (2004). What happens when you give the rewards before the behavior? *Behavior Analysis Digest, 16*(4), 13-14.
16. Renwick, G. (1991). *A fair go for all: Australian/American interactions.* Yarmouth, ME: Nicholas Brealey Publishing.
17. ibidem, p.43.
18. Albom, M. (1997). *Tuesdays with Morrie.* New York: Broadway Books, p.10.
19. Dorman, R. (2006). Lessons for living. In Geller, E.S., & Dean, J. (Eds.). *The power of friendship: Dick Sanderson's positive fight with ALS.* Newport, VA: Make-A-Difference, LLC.

Chapter 5

1. For more on the power of simple conversation see "Coaching is conversing," p. 114, and "Lead with safety conversations," p. 55 in *People-based safety™: The source.*
2. Carnegie, D. (1936). *How to win friends and influence people.* New York: Simon and Schuster; Blanchard, K., & Johnson, S. (1982). *The one-minute manager.* New York: Simon & Schuster, Inc.
3. Krisco, K.H. (1997). *Leadership and the art of conversation.* Rocklin, CA: Prime Publishing.

4. More details are presented on SMART goals in "Set SMART goals," p. 95 in *People-based safety™: The source.*

5. Drucker, P.F. (1999). *Management challenges for the 21st century.* New York: Harper Business.

6. Dean, J. (2006). Blue-collar band of brothers: Construction workers speak up for safety. *Industrial Safety and Hygiene News,* October, p. 40; also posted at www.ishn.com.

7. Covey, S.R. (2004). *The 8th habit: From effectiveness to greatness.* New York: Simon and Schuster, Inc.

8. ibidem, p. 272.

9. ibidem, p. 181.

10. See "Assess levels of trust", p. 281 and "Increase levels of trust", p. 285 in *People-based safety™: The source.*

11. *The American Heritage Dictionary* (1991). Boston, MA: Houghton Mifflin Company.

12. Deming, W.E. (1991, May) *Quality, productivity and competitive position.* Workshop presented by Quality Enhancement Seminars, Inc., Cincinnati, OH.

13. For more on behavioral coaching for safety, see Chapter 3 on "Coaching of people-based safety" in *People-based safety™: The source,* p. 112-167.

14. Rogers, C.R. (1951). *Client-centered therapy.* Boston: Houghton-Mifflin; Rogers, C.R. (1980). *A way of being.* Boston: Houghton-Mifflin.

15. Ellis, A. (1962). *Reason and emotion in psychotherapy.* New York: Kyle Stuart; Ellis, A. (1995). Rational emotion behavior therapy. In R.S. Corsini & D. Wedding (Eds.), *Current psychotherapies* (5th Edition, pp. 162-196). Itasca, IL: Peacock.

16. *The American Heritage Dictionary,* p. 581.

17. ibidem, p. 1046.

18. Festinger, L. (1957). *A theory of cognitive dissonance.* Stanford, CA: Stanford University Press.

19. See "Create internal tension," p. 201 in *People-based safety™: The source.*

20. Geller, E.S. (1991). Don't make safety a priority. *Industrial Safety and Hygiene News,* October, p.12.

21. See "Is safety a priority or a value?" section of "Watch your language" p. 33 in *People-based safety™: The source.*

22. See more details on "the hypocrisy effect" in "Create internal tension", p. 201-203 in *People-based safety™: The source.*

23. Cialdini, R.B. (2001). *Influence: Science and practice* (4th edition). New York: Harper Collins College Publishers.

Chapter 6

1. For more on this connection, see "Thinking and personality," p. 204, *People-based safety™: The source.*

2. Hansen, C.P. (1988). Personality characteristics of the accident involved employee. *Journal of Business and Psychology, 2,* 346-365.

3. See "Traits vs. states," p. 209 in *People-based safety™: The source.*

4. Plomin, R. (1989). Environment and genes: Determinants of behavior. *American Psychologist, 44,* 105-111.

5. Briggs-Myers, I. *Introduction to type.* Palo Alto, CA: Consulting Psychological Press, Inc.

6. Costa, P.T., & McCrae, R.R. (1992). *NEO-PI-R professional manual.* Odessa, FL: Psychological Assessment Resources.

7. Digman, J.M. (1990). Personality structure: Emergence of the five-factor model. *Annual Review of Psychology, 41,* 417-440; John, O.P. (1990). The 'Big Five' factor taxonomy: Dimensions of personality in the natural language and in questionnaires. In Pervin, L.A. (Ed.) *Handbook of personality: Theory and research.* New York: Guilford, p. 66-100.

8. Carey, G., & DiLalla, D.L. (1994). Personality and psychopathology: Genetic perspectives. *Journal of Abnormal Psychology, 103,* 32-43; Loehlin, J.C. (1989). Partitioning environmental and genetic contributions to behavioral development. *American Psychologist, 44,* 1285-1292.

9. Bandura, A. (1997). *Self-efficacy: The exercise of control.* New York: W.H. Freeman and Company.

10. Friedman, M., & Rosenman, R.H. (1974). *Type A behavior and your heart.* New York: Knopf.

11. ibidem

12. For more on the positive attributes of mindfulness, see "The need for mindfulness," p. 168 and "benefits of mindfulness," p. 177 in *People-based safety™: The source.*

13. Taylor, J.A. (1953). A personality scale of manifest anxiety. *Journal of Abnormal Social Psychology, 48,* 285-290.

14. See more details in "Thinking and personality," p. 204 in *People-based safety™: The source.*

15. Covington, M.V. (1992). *Making the grade: A self-worth perspective on motivation and school reform.* Cambridge: Cambridge University Press.

16. Sarason, I.G. (1966). *Personality: An objective approach.* New York: John Wiley & Sons, Inc.

17. Kiersey, D. (1998). *Please understand me II: Temperament, character, intelligence.* Del Mar, CA: Prometheus Nemesis Book Company.

18. True Colors® is a registered trademark of True Colors, Inc., 3605 West MacArthur Boulevard, Suite 702, Santa Ana, CA. (714) 437-5426 or (800) 422-4686, Fax: (866) 374-8958, www.true-colors.com.

19. See "Traits vs. states," p. 209 in *People-based safety™: The source.*

20. Miscisin, M. (2005). *Showing our true colors* (3rd Edition). Santa Ana, CA: True Colors, Inc. Publishing.

21. See "How to give recognition," p. 157 in *People-based safety™: The source.*

22. These suggestions were adapted from the scholarship of Miscisin, M. (2005). *Showing our true colors* (3rd Edition). Santa Ana, CA: True Colors, Inc. Publishing.

23. Gray, J.A. (1982). *The neuropsychology of anxiety: An inquiring into the functions of the septo-hippocampal system.* New York: Oxford University Press; Gray, J.A. (1987). Perspectives on anxiety and impulsivity: A commentary. *Journal of Research in Personality, 21,* 493-509.

24. See "Thinking and personality," p. 204 in *People-based safety™: The source.*

25. Torrubia, R., Avila, C., Molto, J., & Caseras, X. (2001). The sensitivity to punishment and sensitivity to reward questionnaire (SPSRQ) as a measure of Gray's anxiety and impulsivity dimensions. *Personality and Individual Differences, 31,* 837-862.

Index

About the Author

The author, E. Scott Geller, Ph.D., is a Senior Partner of Safety Performance Solutions, Inc., a leading-edge organization specializing in behavior-based safety training and consulting. Dr. Geller and his partners at Safety Performance Solutions (SPS) have helped companies across the country and around the world address the human dynamics of occupational safety through flexible research-founded principles and industry-proven tools. In addition, for almost four decades, Professor E. Scott Geller has taught and conducted research as a faculty member in the Department of Psychology at Virginia Polytechnic Institute and State University, better known as Virginia Tech. In this capacity, he has authored more than 350 research articles and over 75 books or chapters addressing the development and evaluation of behavior-change interventions to improve quality of life.

His recent books in occupational health and safety include: *The Psychology of Safety; Working Safe; Understanding Behavior-Based Safety; Building Successful Safety Teams; Beyond Safety Accountability: How to Increase Personal Responsibility; The Psychology of Safety Handbook; Keys to Behavior-Based Safety from Safety Performance Solutions; The Participation Factor; People-Based Safety™: The Source,* the primer: *What Can Behavior-Based Safety Do For Me?,* and *People-Based Patient Safety™: Enriching Your Culture to Prevent Medical Error,* co-authored by Dave Johnson.

Dr. Geller is a Fellow of the American Psychological Association, the American Psychological Society, and the World Academy of Productivity and Quality Sciences. He is past Editor of the *Journal of Applied Behavior Analysis* (1989-1992), current Associate Editor of *Environment and Behavior* (since 1982), and current Consulting Editor for *Behavior and Social Issues,* the *Behavior Analyst Digest,* and the *Journal of Organizational Behavior Management.*

Scott Geller's caring, dedication, talent, and energy helped him earn a teaching award in 1982 from the American Psychological Association, as well as every university teaching award offered at Virginia Tech. In 1983 he received the Virginia Tech Alumni Teaching Award and was elected to the Virginia Tech Academy of Teaching Excellence; in 1990 he was honored with the all-University Sporn Award for distinguished teaching of freshman-level courses; and in 1999 he was awarded the prestigious W.E. Wine Award for Teaching Excellence.

Dr. Geller has been the Principal Investigator for more than 80 research grants that involving the application of behavioral science for the benefit of corporations, institutions, government agencies, or communities in general. In 2001, Virginia Tech awarded Dr. Geller the University Alumni Award for Excellence in Research. In 2002, the University honored him with the Alumni Outreach Award for his exemplary real-world applications of behavioral science, and in 2003, Dr. Geller received the University Alumni Award for Graduate Student Advising. In 2005, E. Scott Geller was awarded the statewide Virginia Outstanding Faculty Award by the State Council of Higher Education, and Virginia Tech conferred the title of Alumni Distinguished Professor on him. In 2007, Dr. Geller was honored with a lifetime achievement award from the Organizational Behavior Management Network of the International Association of Applied Behavior Analysis.